Heat Transfer 1

Mathematical and Mechanical Engineering Set

coordinated by
Abdelkhalak El Hami

Volume 9

Heat Transfer 1

Conduction

Michel Ledoux
Abdelkhalak El Hami

ISTE

WILEY

First published 2021 in Great Britain and the United States by ISTE Ltd and John Wiley & Sons, Inc.

ISTE Ltd
27-37 St George's Road
London SW19 4EU
UK

www.iste.co.uk

John Wiley & Sons, Inc.
111 River Street
Hoboken, NJ 07030
USA

www.wiley.com

Library of Congress Control Number: 2020949611

British Library Cataloguing-in-Publication Data
A CIP record for this book is available from the British Library
ISBN 978-1-78630-516-9

Contents

Preface

Thermal science is to thermodynamics as decree is to law. It answers the following question – which all good leaders must (or should) ask themselves whenever they have an “idea”: “How would this work in practice?”. In a way, thermal science “implements” thermodynamics, of which it is a branch. A thermodynamics specialist is a kind of energy economist. Applying the first principle, they create a “grocery store”. With the second principle, they talk about the quality of their products. I add or remove heat from a source or work from a system. And the temperature, among other things, defines the quality of the energy for me.

But *by what means do I take or do I give*? Even calculations of elementary reversible transformations do not tell us by what process heat passes from a source to a system.

Thermal science specifies how, but “evacuates” the work. If in a given problem related to, for example, a convector where electrical energy (therefore in the “work” category) appears, it is immediately dissipated into heat by the Joule effect.

Three heat transfer modes can be identified: conduction and radiation – which can be seen separately, although they are often paired up – and convection, which is by nature an interaction of fluid mechanics and conduction.

Dividing the study of thermal science into three is the result of logic. Presenting this work in three volumes is somewhat arbitrary; in our opinion, however, this split was necessary in order to keep the volumes in the collection a reasonable size.

This book is Volume 1 of a collection of problems on heat transfer, devoted to thermal conduction and numerical approaches to such transfers. Despite being a collection of exercises *a priori*, a large part is given over to recalling the practice. To a large extent, the book constitutes a first introduction to the thermal calculation of practical devices, which may be stand-alone. For the subsequent calculations, the reader will not be spared the use of specialist textbooks or encylopedias available in the field of thermal engineering.

The book is intended to reach a wide audience, from technicians to engineers, to researchers in many disciplines, whether physicists or not, who have a one-off transfer problem to resolve in a laboratory context. With this in mind, the theoretical developments in the text itself are as direct as possible. Specialist readers, or those who are simply curious about further theoretical developments (general equations, specific problems, mathematical tools, etc.), may refer to the Appendices.

Volume 1, dedicated to "classic" approaches (analytical treatment) to conduction, will be of interest primarily to readers who are looking for "simple" prediction methods.

After the generalities outlined in **Chapter 1**, **Chapter 2** presents the physical laws of conduction, describing the Fourier law and setting out the heat equation. We then pinpoint the fundamental content of the problems found in conduction and their approach. At this point, we will need to distinguish between stationary (we also refer to permanent regimes) and non-stationary problems.

Chapter 3 deals with conduction in a stationary regime. Emphasis is logically placed on plane and cylindrical geometries. Importance is placed on the concept of thermal resistance, an essential tool for thermal scientists, in both of these geometries. This chapter provides many examples of application of this concept.

Chapter 4 presents the notion of a quasi-stationary regime; this method, although approximate, does in fact have undeniable practical scope. Valid for relatively slow transfers – presuming the temperature environment is homogeneous – instantaneous balances can be obtained, in addition to the laws of thermal evolution that allow valid approximations to the problems that, in principle, relate to the variable regime. Again, in this context, this chapter deals with the plane and cylindrical problems.

Chapter 5 deals with variable conduction regimes. The most classic monodimensional problems are tackled: fixed temperature at the interface at the instant $t=0$, non-stationary conduction at constant flow density, temperature with sinusoidal variation fixed at the interface and the problem of two adjoining walls.

Many examples are provided to illustrate this important aspect of conductive transfers.

Chapter 6 presents the notion of fin theory, which is associated with a few simple examples.

Within the **Appendices**, there are tables of error functions and their offshoots: erf, erfc, ierfc, as well as reminders that are often essential for hyperbolic functions. They also provide information on the notion of treating certain non-stationary problems using Laplace transformations. Again, many examples are given to illustrate another important aspect of conductive transfers.

November 2020

Introduction

I.1. Preamble

Thermal energy was probably first perceived (if not identified) by humanity, through the Sun. The themes of night and day are found at the center of most ancient myths. Humanity's greatest fear was probably that the Sun would not return again in the morning. Fire became controlled in approximately 400,000 BP. Thermal transfer was therefore a companion of *Homo ergaster*, long before *Homo sapiens sapiens*.

However, it took a few hundred thousand years before so-called "modern" science was born. Newtonian mechanics dates from three centuries ago. Paradoxically, another century and a half passed by before energy was correctly perceived by scientists, in terms of the new field of thermodynamics. Furthermore, a systematic study of heat transfer mechanisms was carried out at the end of the 19th century, and even later for the study of limit layers, the basis of convection.

Heating, lighting and operating the steam engines of the 19th century were all very prosaic concerns. Yet this is where revolutions in the history of physics began: the explosion of statistical thermodynamics driven by Boltzmann's genius, and quantum mechanics erupted with Planck, again with Boltzmann's invovlement.

Advances in radiation science, particularly in sensor technology, have enabled us to push back our "vision" of the universe by a considerable number of light years. To these advances we owe, in particular, the renewed interest in general relativity that quantum mechanics had slightly eclipsed, through demonstration of *black holes*, the physics of which may still hold further surprises for us.

Closer to home, fundamental thermal science, whether it is conduction, convection or radiation, contributes to the improvement of our daily lives. This is particularly true in the field of housing where it contributes, under pressure from environmental questions, to the evolution of new concepts such as the active house.

The physics that we describe in this way, and to which we will perhaps introduce some readers, is therefore related both to the pinnacles of knowledge and the banality of our daily lives. Modestly, we will place our ambition in this latter area.

There are numerous heat transfer textbooks in different formats: “handbooks” attempting to be exhaustive are an irreplaceable collection of correlations. High-level courses, at universities or engineering schools, are also quite exhaustive, but they remain demanding for the listener or the reader. Specialist, more empirical thematic manuals are still focused on specialists in spite of all this.

So why do we need another book?

The authors have taught at university level and in prestigious French engineering schools, and have been involved in the training of engineers on block-release courses. This last method of teaching, which has been gaining popularity in recent years, particularly in Europe, incorporates a distinctive feature from an educational point of view. Its practice has, in part, inspired this book.

The aim is to help learners who have not had high-level mathematical training in their first years following the French Baccalaureate (therefore accessible to apprentices), and pupils with more traditional profiles. At the same time, we would like to show this broad audience the very new possibilities in the field of digital processing of complex problems.

When a miner wants to detach a block of coal or precious mineral from a wall, they pick up a pneumatic drill. If we want to construct a tunnel, we must use dynamite. The same is true for physicists.

Whether they are researchers, engineers or simply teachers, scientists have two tools in their hands: a *calculator* and a *computer* (with very variable power). Since both authors are teacher researchers, they know they owe everything to the invention of the computer. From the point of view of teaching, however, each one of the two authors has remained specialist, one holding out for the calculator and "back-of-the-napkin" calculations, and the other one using digital calculations.

The revolution that digital tools has generated in the world of "science" and "technical" fields, aside from the context of our daily lives, no longer needs to be proved. We are a "has been" nowadays if we do not talk about Industry 4.0. The "digital divide" is bigger than the social divide, unless it is part of it....

Indeed, the memory of this revolution is now fading. Have students today ever had a "slide rule" in their hands? Do they even know what it is? Yet, all the physicists behind the laws of thermal science had only this tool in hand, giving three significant figures (four with good visibility and tenacity), leaving the user to find the power of ten of the result. It goes without saying that a simple calculation of a reversible adiabatic expansion became an ordeal, which played a part in degrading the already negative image of thermodynamics held by the average student.

This reminder will seem useless to some; slide rules are at best sleeping in drawers. But there is a moral to this story: no matter what type of keyboard we type on, a calculator or a computer, our head must have control over our fingers. This book has been written on the basis of this moral.

A good physicist must have a perfect understanding of the idea of an "order of magnitude". For this, the tool is a calculator. We always do a rough sizing of a project before moving on to detailed modeling and numerical calculations.

The two authors belong to the world of engineering sciences, meaning most of their PhD students have entered the private sector. One of them, having moved into the aerospace sector, came back to see us very surprised by the recurrence of "back-of-the-napkin" calculations in his day-to-day work.

Fundamental or "basic physics" concepts are taken from a type of manual that is resolutely different from those dedicated to the numerical approach. In this case, the authors allow themselves to believe that it is no bad thing to collect them all together in a single book, for once. This is a significant difference that will surprise some and, without doubt, be criticized by others. Nevertheless, when reading this book, an "average" student will be initiated to a field that teaching models generally promised "for later on" (or never if he/she never goes beyond a certain level of education). It is also true that fully immersed in equations and complex calculations, specialist readers will be able to "be refreshed" when faced with the short exercises, which can sometimes surprise and encourage them – why not – to go back to their roots (assuming they had indeed been there).

Another significant difference is that this book is directed at a large scientific audience, which covers possibly the entire field: researchers, PhD students or those who have obtained Confirmation and are just starting out in the field, technicians, students or professionals, engineers. This last type of scientist is perhaps the main target of this book.

So, what is this book for?

Above all, it **contains problems to be worked on**, of which most are accessible to all, from the level of an apprentice technician upwards, either one or two years after the Baccalaureate. This book was written in France, where scientific teaching is structured around universities, engineering Grandes Écoles, engineering training through apprenticeships and two types of technician training sections at high schools or universities. In countries with simpler models, readers should also find it useful.

It seems necessary to surround these problems with strong reminders of past learning, so that the reader does not need to permanently refer back to their manuals. We see two advantages in this: a presentation of the scientific material focusing on the problems, and a second chance for readers to integrate notions that perhaps had not been well understood in the initial teaching.

Lastly, upon rereading, the authors also recommend this book as an introduction to the taught disciplines.

I.2. Introduction

Thermal science is to thermodynamics as decree means is to law. It answers the following question – which all good leaders must (or should) ask themselves whenever they have an “idea”: “How would this work in practice?”.

In a way, thermal science “implements” thermodynamics, of which it is a branch.

A thermodynamics specialist is a kind of energy economist. Applying the first principle, they create a “grocery store”. With the second principle, they talk about the quality of their products. I add or remove heat from a source or work from a system. And the temperature, among other things, defines the quality of the energy for me.

But *by what means do I take or do I give*? Even calculations of elementary reversible transformations do not tell us by what process heat passes from a source to a system.

Thermal science specifies how, but “evacuates” the work. If in a given problem related to, for example, a convector where electrical energy (therefore in the “work” category) appears, it is immediately dissipated into heat by the Joule effect.

Three heat transfer modes can be identified: conduction and radiation – which can be seen separately, although they are often paired up – and convection, which is by nature an interaction of fluid mechanics and conduction.

Dividing the study of thermal science into three volumes is the result of logic. Presenting this work in three volumes is somewhat arbitrary; in our opinion, however, this split was necessary in order to keep the volumes in the collection a reasonable size.

The first volume, entitled **Heat Transfer 1**, is dedicated to “classic” approaches (analytical treatment) to conduction, which will be of greater interest to readers who are looking for “simple” prediction methods.

The second volume, entitled **Heat Transfer 2**, is dedicated to “classic” approaches (analytical treatment) of radiation, and assembles digital approaches of these various transfer modes. It is aimed at engineers or researchers who want to resolve more complex problems.

The third volume, entitled **Heat Transfer 3**, is focused on convection transfers. As we have already pointed out, all of these transport operations are rarely pure and lead to problems that involve three inter-connecting transfer modes, conduction, convection and radiation.

Before our readers immerse themselves in a text that, despite our best efforts, remains intellectually demanding, we propose a short text that is a little lighter.

I.3. Interlude

Let us imagine, in a "B movie" context, a somber hostel in the gray fog of a port in the middle of nowhere. Sailors from a faraway marina come and drink away their troubles. And as always, the drink helping them along, they turn to fighting.

Let us entrust Ludwig Boltzmann to direct the film. Our B movie heroes are getting agitated, delivering blows to one another. Each one of them has moderate kinetic energy, distributed heterogeneously among them in the room. For some reason, they get involved in a general brawl. Their average kinetic energy becomes much greater.

In everyday language, we would say that **things are hotting up**.

This would bring us right into line with a fundamental concept of Boltzmann, who was the first to hypothesize that **heat is made up of molecular agitation**. *The temperature in a gas is proportional to the average quadratic energy of the molecules that make it up*:

$$E_C = \frac{1}{2} k T$$

Using this model, we will return to the physical basis for all transport phenomena.

On the way, we rarely escape from the explosion of a door or a window, giving in under the repeated beatings of the brawlers.

We have just modeled the pressure, due to the transfer of the quantity of movement on the surface, by the impact of molecules.

Let us now imagine that the altercation is initially located in the corner of the room: a smaller group starts fighting between themselves.

From kicks to punches, after multiple impacts within the group and its immediate neighbors, the agitation will spread: *we have just seen the mechanism of heat propagation by transfer of impacts.*

Let us place an imaginary separation (geometrically but immaterially defined) at the center of the room. Let us count the sailors that cross through it within a unit of time.

This wall is now crossed by kinetic energy: we have defined a flow of heat.

Let us put a metal ring with a surface area of $S = 1\,m^2$ in the room. On both sides of this ring, the blows exchanged constitute a transfer of kinetic energy – we have just defined **the heat flow density**.

And we have just understood the nature of the propagation of thermal flows by impacts.

Let us suppose that the great majority of the brawlers come from a ship with a white uniform. Let us suppose that another boat in the port has uniforms that are red. The red ones are initially all united. We will then quickly see that the red mariners, as they receive and return blows, spread out across the room. *We have just shown the mechanism of* **diffusion of matter**, *of a component within a mixture.*

We will have a better qualitative understanding that the fundamental law of conduction (Fourier Law) is formally identical to the law for the diffusion of mass (Fick Law).

Let us put our agitated sailors in the compartments of a flatcar train, where they continue to fight. And let us start the train moving. The kinetic energy that they contain is transported from one point to another.

We have just invented **thermal convection**.

We can go further.

Let us imagine a series of flatcar trains on a set of parallel tracks. The train furthest to the side is fixed to a platform. All of these trains are full of sailors. Let us suppose that our train follows the outside, parallel rail tracks. No brakes will prevent these trains from moving. Only the last train, at the platform, is stuck.

For a reason we do not need to analyze (cinema allows all kinds of fantasy), "clusters" of fighting sailors jump from one wagon to the next. These "clusters" contain a component of speed that is parallel to the train, which will communicate information about the quantity of movement to the adjacent train. These trains will then start to move, more quickly the closer they are to the outside train. And the same occurs up to the train at the platform. This train will not move, but a force will be applied to its brakes.

We have just discovered the mechanism of dynamic viscosity. At the same time, the parallel trains in relative movement give us a picture of the *notion of boundary layers.*

At the same time, these agitated clusters carry their disordered kinetic energy, "thermal" agitation. We have just seen the mechanism of the *thermal boundary layer.*

Finally, let us include a few red mariners in the crowd of white. They will be carried with the clusters, and we have just invented *the limit layer of diffusion of a species.*

We are in a fantasy, and let us benefit from it as far as we can. To finish, let us suppose that this is carnival day; each sailor has a belt equipped with bells.

All the individuals have a different speed, and the impacts are random, all the bells start to jingle, each with a different frequency. The distribution of frequencies will depend on the statistical distribution of speeds (Boltzmann statistics), and the intensity of noise produced will depend on the total agitation energy of the sailors.

We have just understood the **basic mechanism of radiation**. *We have just realized why the theory of radiation needed to use the concepts of* *statistics derived from the work of Boltzmann – a brilliant pupil of Planck – to produce the emissions spectrum of a black body, for example.*

NOTE.– *the model is certainly simplistic. The emission comes from quantum transitions in the gas atoms.*

Here, we have already deviated from the pure substance of the book, but we could go even further.

Let us suppose that our agitated sailors are in a room with one mobile wall (a nightmare scenario frequently seen on the silver screen).

The incessant impacts of the fighters on this wall create a force that pushes it. This force, reduced to a surface unit, explains the notion of pressure.

By pushing against this wall, our crowd applies work that is greater than the resistance.

Here we see an equivalence spring up between work and heat that, at a fundamental level, are simply two mechanical energies: one ordered and the other disordered. **The first principle of thermodynamics is illustrated by this.**

We can see that the incidence of an average blow on the wall is rarely normal.

Therefore, an average fighter will have a trajectory that will be reflected off the wall. And only the normal component of its speed will be able to push (or transfer work to) the wall.

Thus, we see that it will be impossible for the crowd (taken to mean a gas) to give all its energy to a mobile wall.

The fundamental mechanism that leads to the **second principle of thermodynamics** *has just been demonstrated.*

These “light-hearted” images, which will perhaps not please everyone, were an oral support for the presentation of different transport phenomena by one of the authors. We hope that the reader, once they have studied this book, will want to return to this text. They will then have understood, we hope, the images that lead to the development of thermodynamics.

And if this text has a moral, it would be: *Writing down thermodynamics, just like thermal science, is based on continuous equations. The fundamentals of physics that determine these phenomena arise from the field of the discontinuous: discontinuity of matter, divided into particles; discontinuity of light, divided into photons.*

1

The Problem of Thermal Conduction: General Comments

1.1. The fundamental problem of thermal conduction

The fundamental problem of thermal conduction involves the determination of temperature domains and flows across particular surfaces, for a given physical situation, in one or several given environments.

The resolution of a thermal conduction problem involves:

a) for all problems, a heat equation is considered, resulting from a local heat balance;

b) for specific conditions of the problem, the conditions are constituted at the limits that are applied to the heat equation.

In the most general case, these temperature domains T are not homogeneous. They can be three-dimensional and variable over time:

$$T = T(x, y, z, t)$$

In other systems, we note that $T = T(r, \theta, z, t)$ or $T = T(r, \theta, \Phi, t)$.

Temperature is expressed in degrees Celsius (°C), formerly degrees centigrade, or in Kelvin (K).

We know that the temperature, expressed in Kelvin, is measured from absolute zero. The conversion rule is known as: $T\ K = T\ C + 273.15$.

The most general problem presented in thermal conduction is therefore extremely complex, since the heat equation has four dimensions (three in space and one in time).

Fortunately, significant simplifications are possible for many problems.

The temperature field can only depend on a spatial variable. We say that conduction is monodimensional or bidimensional.

The temperature can remain fixed at each point as time progresses.

This last remark divides the approach that we will adopt into two parts:

– **stationary** conductive heat transfer;

– **non-stationary** conductive heat transfer.

Two important categories will be examined in this chapter:

– problems of stationary conduction with a single dimension: $T = T(x)$ or $T = T(r)$

– problems of non-stationary conduction with a single dimension: $T = T(x,t)$ or $T = T(r,t)$

For many problems, the analytical approach is possible. This will be the case, in particular, for the stationary or non-stationary problems with a single dimension. In (most) other cases, a numerical approach is required.

1.2. Definitions

1.2.1. *Temperature, isothermal surface and gradient*

The temperature T is a parameter defined in all thermodynamics classes.

As of now, we can define **isothermal surfaces** in space. An isothermal surface is a surface on which the temperature is constant:

$$\text{Isothermal surface } S \quad \Rightarrow \quad T = Cte \text{ for an entire surface.} \qquad [1.1]$$

Furthermore, we will look at the important relationship between this surface and heat flow.

Let us note that in a non-stationary problem, the isothermal surface is mobile in time.

The temperature field also allows a very important vector to be defined, whose use will be explained later on: ***temperature gradient***.

Initially, we define this vector in a Cartesian system. Later on, we will see that it is quite easy to determine its components in other systems when the symmetries in a problem make that a pertinent option.

In a Cartesian system, this gradient is constructed by taking, for each of its components, the partial derivative of the temperature with respect to the relevant coordinate axis. In other words:

$$\overrightarrow{grad}\,T \left| \begin{array}{l} \dfrac{\partial T}{\partial x} \\ \dfrac{\partial T}{\partial y} \\ \dfrac{\partial T}{\partial z} \end{array} \right. \qquad [1.2]$$

For other systems, it will be useful to refer to **Appendix 2**.

This mathematical operation is known to all physicists. With a minus sign, it is the derivation of a force from a potential.

We should note that in a stationary problem, this gradient is a vector attached to each point in space and fixed in time. In the case of a non-stationary problem, this gradient is a vector attached to each point in space and variable in time.

The temperature gradient possesses a fundamental property, which is demonstrated in Appendix 2: the gradient vector is normal at each point in space to the isothermal surface that passes through this point.

This property will be familiar to those who have studied electric fields.

1.2.2. *Flow and density of flow*

These two essential parameters will be used throughout this book.

Thermal flow Φ is the quantity of heat Q that passes a given surface in a unit of time:

$$\Phi = \frac{Q}{t} \qquad [1.3]$$

This flow is expressed in Joules per second ($J.s^{-1}$) or even better, in Watts (W).

Density of the thermal flow φ, which can be even greater, is the flow that passes through a unit of surface:

$$\varphi = \frac{\Phi}{S} = \frac{Q}{tS} \qquad [1.4]$$

This flow density is expressed in Watt per m² ($W.m^{-2}$).

It is important not to confuse these two parameters, which also do not have the same units.

In this book, we will systematically note the flow with a capital Φ and the flow densities with a lowercase φ.

These flows and flow densities may not be uniform through a surface, which can also vary rapidly over time. We then deal with a definition on a differential basis, through small surface areas and with small time intervals.

Through a small surface dS:

$$d\Phi(t) = \frac{dQ}{dt} \qquad [1.5]$$

$$\varphi = \frac{d\Phi}{dS} = \frac{d^2Q}{dt\,dS} \qquad [1.6]$$

The quantity of heat d^2Q is measured on a small surface $d\,S$ and on a small time interval $d\,t$, which becomes twice differential. The flow $d\,\Phi\,(t)$ becomes differential since it is the relationship of a twice differential quantity by a small surface $d\,S$ at a given time t. The flow density φ is the relationship between two differential quantities, which remains finite (expressed in flow per unit of surface and time).

1.3. Relation to thermodynamics

We have already pointed out that when we talk about thermal science, thermodynamics is not far off. More precisely, it is its main principle.

Before the first principle, there was calorimetry.

We should be aware that each time that we draw up a balance for an environment, which will be particularly true in **Chapter 4**, we link a variation in temperature to an exchange of heat by a calorimetric operation.

1.3.1. *Calorimetry*

Calorimetry was, in fact, born before the first principle. We (mainly chemists) noted that the relative variations in temperature due to an exchange of what we had then designated as "heat", depended on the mass m of the environments and a coefficient that is today denoted as the mass calorific capacity c. In other words, when a body transfers heat to another body, the relationship (with an obvious notation) is:

$$m_1\,c_1\,\Delta T_1 = -m_2\,c_2\,\Delta T_2 \qquad [1.7]$$

The minus sign indicates that one body is cooling and the other is heating.

This leads to the relationship between the quantity of heat Q given to a body and its rise in temperature ΔT (or the inverse in the case of cooling), which can be written as:

$$Q = m\,c\,\Delta T \qquad [1.8]$$

1.3.2. *The first principle*

Later on, Joule will demonstrate the equivalence of two forms of energy: ***work and heat***.

His experiment can be reproduced on a daily basis. It is sufficient to turn a small paint mixer in water contained in a Thermos flask. Why a Thermos? So that the heat given to the water does not escape.

We measure the electrical energy provided to an engine for a certain time, with energy measuring the ***work "given"*** to the water. The heating of the water whose mass and calorific capacity are known measures the ***quantity of heat brought*** to the water. This quantity of heat can only come from the transformation of work provided through heat. In Joule's era, heat was measured in calories.

One calorie raised the temperature of 1 gram of water by 1 degree Celsius. Work is evaluated in Joules. Modestly, this author only had a kilogram meter.

Hence, two conclusions will always be in the background of our study:

– qualitatively: work and heat are two equivalent and interchangeable forms of energy;

– quantitatively: the equivalence between a calorie and (today) a Joule is written as:

$$1\,calorie = 4.18\,Joules \qquad [1.9]$$

In principle, this is part of the repertoire of all high school pupils, but it is not a bad thing to give a reminder of it.

1.3.3. *The second principle*

This principle is about quality and reversibility. It is easy to transform work into heat. Simple friction is sufficient. This is what Joule experienced.

It is more complicated to transform heat into work. Several sources are required, at least one hot and one cold, and the temperature of the sources becomes important. Moreover, we never recover all the heat that we provide with the hot source in work. We always "throw" some into the cold source. This will be the large Carnot work, among others.

Those who have closely read the text at the end of our preamble will understand why. The blades of the mixer transform a "directed" energy into a chaotic thermal agitation. Chaos will also never spontaneously return to order.

All this comes from the genius (we defend our choice of word) of Boltzmann, who will also be mentioned several times in the following chapters.

To conclude, let us say that in all our transfers, we are dipping into the realm of irreversibility and the permanent creation of entropy, but this will be self-evident for the reader.

2

The Physics of Conduction

2.1. Introduction

As we have known from time immemorial, from a qualitative point of view, heat only moves "spontaneously" from hot(ter) zones to cold(er) zones. This is the phenomenon of conduction. In thermodynamics, spontaneously means without the intervention of "work".

Thermodynamics reveals that heat is made up of the kinetic energy of the constituents of matter, molecules or atoms that move with a certain agitation. The temperature has indeed been linked to the average kinetic energy of these molecules by Boltzmann. The molecules are immobile at absolute zero, which therefore becomes the minimum physically accessible temperature.

From a quantitative point of view, transfers by conduction are governed by an apparently simple general rule: Fourier's law.

2.2. Fourier's law

2.2.1. *Experiment*

In order to set up this law, we imagine a simple device, which is represented in Figure 2.1.

A material of thickness e and cross-section S is placed between two isothermal plates. The temperatures T_1 and T_2 of the left-hand and right-hand faces of the test material are measured using thermocouples, for example. The left-hand plate is equipped with an electrical resistance that delivers a power of $P = \dfrac{V^2}{R}$. The right-hand plate is cooled.

The whole apparatus is placed within a very thick layer of material that is a very poor conductor of heat. Here, we make the hypothesis that the heat dissipated by the electrical resistance passes all the way through the material by thermal conduction.

The density of thermal flow is constant in the material, from left to right, and is equal to:

$$\varphi = \frac{P}{S} \qquad [2.1]$$

We therefore observe, after many tests (in particular, with variation of P and e), that for a given material, the flow density is related to the thickness and temperature by:

$$\varphi = \lambda \frac{T_1 - T_2}{e} \qquad [2.2]$$

where λ is a characteristic coefficient of the material being tested.

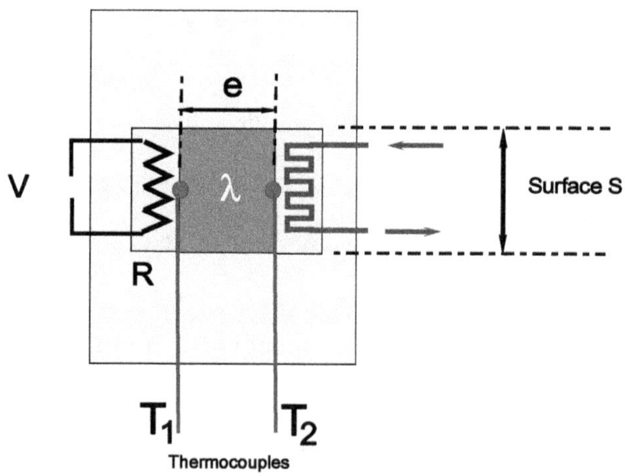

Figure 2.1. *Diagram of a simplified device for determining Fourier's law. For a color version of this figure, see www.iste.co.uk/ledoux/heat1.zip*

This coefficient, a property of the material, is known as the **thermal conductibility of the material**. It is expressed in $J\,s^{-1}m^{-1}K^{-1}$ or much more frequently in $W\,s^{-1}K^{-1}$

In many situations, we can consider this conductibility to be constant. The experiment demonstrates that this conductibility depends on the temperature. In certain problems, in particular in the case of high temperatures and heterogeneous thermal fields, we must take into account the spatial variation of the thermal conductibility within the material, which can be expressed explicitly as:

$$\lambda = \lambda(x, y, z) \qquad [2.3]$$

or more often by a relationship between the conductibility and the temperature:

$$\lambda = \lambda(T) \qquad [2.4]$$

Materials with high values of λ are known as conductors. Materials with low values of λ are known as insulators.

Here are a few values to set out the general idea:

Metals are conductors of heat:

$$\lambda_{Copper} = 386\ W\,s^{-1}K^{-1}$$

$$\lambda_{Al} = 204\ W\,s^{-1}K^{-1}$$

Many solids (including refractory ones) are insulators:

For example, for a light resinous wood:

$$\lambda\ = 0.12\ W\,s^{-1}K^{-1}$$

Liquids are relatively poor conductors of heat. For example, for water at 20°C:

$$\lambda_{Water} = 0.597\ W\,s^{-1}K^{-1}$$

Gases are poor conductors of heat. For example, for air at $20°C$:

$$\lambda_{air} = 0.0257\ W\, s^{-1} K^{-1}$$

REMARK.– Air is thus one of the best thermal insulators. However, between two faces, separated by a layer of air of thickness of the order of 2 cm, natural convection phenomena leads to a "parasitic" thermal transfer. Our solution is thus to trap air in alveolar structures, such as networks of mineral threads (glass fiber or rock "wool"). The solid material then leads to parasitic thermal conduction, known as a "thermal bridge", which explains how *rock wool* presents a thermal conductibility of the order of $\lambda = 0.4\ W\, s^{-1} K^{-1}$, higher than that of air. If the faces are made of glass (double glazing), the distance between the two panes must be small.

We can also observe that strictly speaking, a vacuum is an absolute insulator, since there are no molecules that are likely to propagate impacts. We will see later on however, that a vacuum is ideal for the propagation of radiation and in that sense does not present an obstacle to the transmission of heat. This explains the structure of a thermos flask, whose internal walls, positioned face to face, are made of reflective glass.

The metrology described here is used in practice. It turns out to be difficult to use, insofar as any three-dimensionality of heat flows is a source of error. It therefore requires the use of a material that is a better insulator than the one being tested, or failing this, very large volumes.

Other "dynamic" methods using non-stationary thermal transfer properties (impulse response methods) have been developed.

2.2.2. *Temperature profile*

The law $\varphi = \lambda \dfrac{T_1 - T_2}{e}$ implies that, in the case of the experiment, in other words, a mono-dimensional case, the profile of speed $T(x)$ is linear. Here, we have defined x as the x-axis of a plane that is parallel to the walls of the material (plates) located on an axis Ox, which is perpendicular to the heating and cooling plates. $T(x)$ is then the temperature of a plane of the material, perpendicular to Ox.

Indeed, the law $\varphi = \lambda \frac{T_1 - T_2}{e}$ is valid for all distances e between two planes perpendicular to Ox. This remains true for two planes that are extremely close together. $e = dx$ is then an infinitely small distance (as small as we want), and the difference in temperature $T_1 - T_2$ becomes a very small difference dT. Fourier's law is then written, in differential form, as:

$$\varphi = \lambda \frac{dT}{dx} \qquad [2.5]$$

$e = dx$ is as small as we want, and $\frac{dT}{dx}$ is then the derivative of $T(x)$ with respect to x.

φ is constant, so the function $T(x)$ is linear.

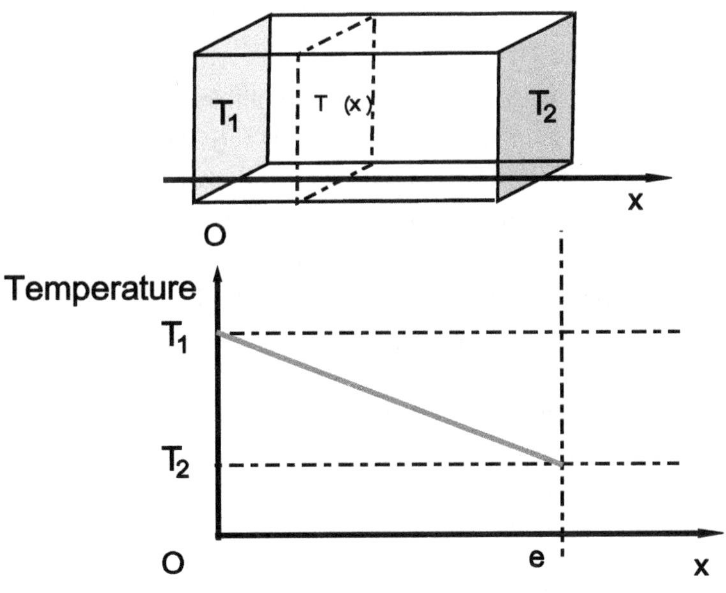

Figure 2.2. *Linear temperature profile in a homogeneous material. For a color version of this figure, see www.iste.co.uk/ledoux/heat1.zip*

2.2.3. *General expression of the Fourier law*

The above explanations are sufficient for mono-dimensional problems. The general heat equation implies extending this law to the three-dimensional case. To do so, we will use the following reasoning. A reminder of this calculation will be given in the Appendix.

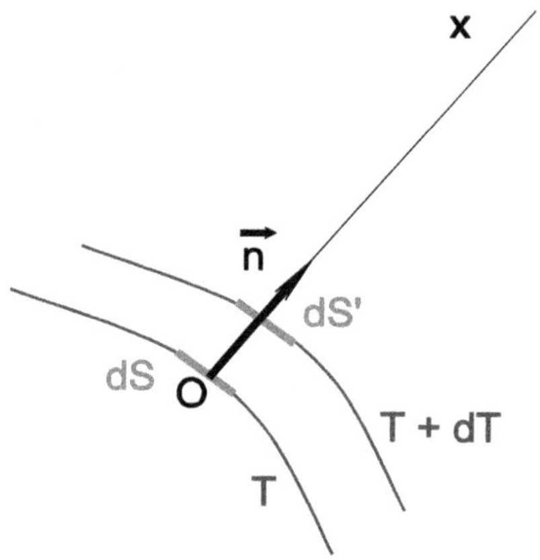

Figure 2.3. *Case of two infinitely close isotherms; relationship with the gradient. For a color version of this figure, see www.iste.co.uk/ledoux/heat1.zip*

Let us consider two isothermal surfaces with respective temperatures T and $T+dT$, positioned very close to each other. In other words, M, a point on isotherm T. We will now consider two infinitely small surfaces $d\,S_1$ and $d\,S_2$, neighbors on each of the isotherms, at point M. They have practically the same normal $\vec{n}$, directed conventionally from the surface with temperature T towards the surface of $T+dT$. Locally, we observe the situation found in a plane mono-dimensional system.

The thermal flow that traverses $d\,S_1$ and $d\,S_2$ will be expressed as:

$$d\,\Phi = -\lambda \frac{d\,T}{d\,x} d\,S_1 \qquad [2.6]$$

where x is an axis counted on the axis for $\vec{n}$. Let us recall that the minus sign indicates that heat propagates in the direction of decreasing temperatures. In this expression, the sign implies a positive flow when it flows in the direction of the normal.

Let us consider a surface $d\,S'$ based on the same section of cylinder as $d\,S_1$ and $d\,S_2$, whose normal $\vec{n}'$ makes an angle θ with the normal $\vec{n}$ that is common to $d\,S_1$ and $d\,S_2$.

The same thermal heat flow evidently traverses $d\,S'$, $d\,S_1$ and $d\,S_2$

We have a known geometrical relationship:

$$d\,S_1 = d\,S'\cos\theta \qquad [2.7]$$

And the traversing flow $d\,S'$ is written as:

$$d\,\Phi = -\lambda\frac{d\,T}{d\,x}d\,S_1 == -\lambda\frac{d\,T}{d\,x}d\,S'\cos\theta \qquad [2.8]$$

Near M, the temperature gradient is reduced to a component along the axis of x:

$$\left(gra\vec{d}\,T\right)_x = \lambda\frac{d\,T}{d\,x} \qquad [2.9]$$

which we can re-write as:

$$\left\|gra\vec{d}\,T\right\| = \left|\lambda\frac{d\,T}{d\,x}\right| \qquad [2.10]$$

We note that the normal for a unit vector is one:

$$\left\|\vec{n}'\right\| = 1 \qquad [2.11]$$

In fact, the flow will take the form of a scalar product:

$$d\,\Phi = -\lambda\frac{d\,T}{d\,x}d\,S'\cos\theta = \left\|gra\vec{d}\,T\right\|\left\|\vec{n}'\right\|\cos\theta\,d\,S' \qquad [2.12]$$

$$d\,\Phi = \left\|gra\vec{d}\,T\right\|\left\|\vec{n}'\right\|\cos\left(gra\vec{d}\,T,\vec{n}'\right)d\,S' \qquad [2.13]$$

Taking into account the orientations of the normal described above, we obtain the general form of the Fourier law:

$$d\Phi = -\overrightarrow{grad}\,T.\vec{n}'\,dS' \qquad [2.14]$$

REMARK.– Thus, the thermal flow that traverses a closed surface will be written as:

$$\Phi = \iint_S \lambda\,\overrightarrow{grad}\,T.\vec{n}\,dS \qquad [2.15]$$

2.3. Heat equation

2.3.1. *General problem*

All thermal conduction problems go through several stages, for any selected approach:

a) setting out a thermal balance. This balance can be written explicitly for each problem, and this will be the case, in particular, in **Chapter 4** concerning quasi-stationary regimes.

We can also use an assessment written once and for all for a small region: this will be the heat equation;

b) writing special conditions for the problem, which will constitute the conditions at the limits of the equation. This point will be examined in greater detail in **section 2.4**, which follows.

This paragraph is dedicated to establishing the heat equation.

To describe the most general problem, this equation is a partial differential equation whose solution is the local and instantaneous temperature, in other words a variable with four dimensions: three spatial variables and one time variable. We could, depending on the symmetries of the problem in question, work in different coordinate systems. Throughout the problems posed in this book, we will frequently work in Cartesian coordinates (problems known as "plane problems"), in cylindrical coordinates, or more rarely, in spherical coordinates.

The temperature will therefore be expressed in different ways:

– Cartesian coordinates:

$$T = T(x, y, z, t) \quad [2.16a]$$

– cylindrical coordinates:

$$T = T(r, \theta, z, t) \quad [2.16b]$$

– spherical coordinates:

$$T = T(r, \theta, \Phi, t) \quad [2.16c]$$

A general problem of this kind is expressed in four dimensions; it is one of the most complex problems to deal with.

Often the device in question has symmetries that allow the problem to be reduced to one spatial dimension. The temperature now only depends on two variables, space and time. We can distinguish between:

– plane problems:

$$T = T(x, t) \quad [2.17a]$$

– cylindrical symmetry:

$$T = T(r, t) \quad [2.17b]$$

– spherical symmetry:

$$T = T(r, t) \quad [2.17c]$$

In these cases, the heat equation, becoming an equation of two dimensions, will take on a simplified form.

The problems dealt with in this guide will almost systematically be related to this single-dimensional hypothesis.

For this reason, we will initially establish the heat equation in two dimensions. This construction will be based on a direct expression of the thermal balance.

As a matter of general culture, we consider it important for thermal scientists to know the full equation, which is often a starting point for various scientific publications that they will need to consult over the course of their practice.

For this purpose, we will establish the heat equation in four dimensions, using the tools of vector geometry. For readers who are not accustomed to these methods and to lighten the explanations a little, we have put this establishment in the Appendix, simply giving its result below.

2.3.2. *Mono-dimensional plane problem*

2.3.2.1. *The terms of conduction per m² of wall*

The problem is mono-dimensional: T only depends on x. Therefore, any plane perpendicular to Ox is the isotherm $T = T(x)$.

In other words, an axis Ox which is used as a reference framework for the direction of heat propagation; this axis is orientated from left to right in Figure 2.4.

We write the instantaneous heat balance for a given volume determined by two planes normal to Ox, infinite neighbors of axes x and $x + ds$ and surface S with a unit surface area for simplification purposes: The flows that traverse it are therefore equal to the densities of flow ϕ.

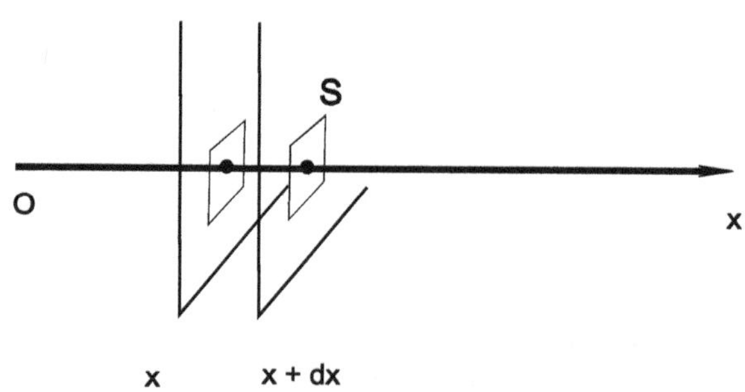

Figure 2.4. *Establishment of the mono-dimensional heat equation*

In other words, the axis plane x, perpendicular to Ox; let us consider the quantity of heat that traverses it per m², or density of flow φ:

Let us write

$$\varphi(x) = -\lambda \frac{dT}{dx} \qquad [2.18]$$

What is the role of the minus sign in this expression?

$\frac{dT}{dx} < 0$ when the heat moves from left to right, and $-\lambda \frac{dT}{dx}$ is positive when the heat moves in the direction of increasing x.

$\frac{dT}{dx} > 0$ when the heat moves from right to left, and $-\lambda \frac{dT}{dx}$ is negative when the heat moves in the direction of increasing x.

The expression $-\lambda \frac{dT}{dx}$ is written at point x of the axis Ox. How is $-\lambda \frac{dT}{dx}$ written when we go from the x-axis plane to the $x + dx$-axis plane, with dx infinitely small?

General case of any function $f(x)$: when x goes from x to $x + dx$, a function $f(x)$ goes from $f(x)$ to $f(x) + \frac{\partial f}{\partial x} dx$.

$\varphi(x) = -\lambda \frac{dT}{dx}$ is a specific $f(x)$ function.

Therefore, when we go from x to $x + dx$, $-\lambda \frac{dT}{dx}$ becomes

$$\varphi(x) + \frac{d\varphi(x)}{dx} = -\lambda \frac{dT}{dx} + \frac{d}{dx} \quad -\lambda \frac{dT}{dx} \quad dx \qquad [2.19]$$

Let us now consider a plane located at $x+dx$ and the quantity of heat that traverses it per m².

We write:

$$\varphi(x+dx)=-\lambda\frac{dT}{dx}+\frac{d}{dx}\quad-\lambda\frac{dT}{dx}\quad dx \qquad [2.20]$$

Using the same reasoning made for x,

$-\lambda\frac{dT}{dx}+\frac{d}{dx}\left(-\lambda\frac{dT}{dx}\right)dx$ is positive when the heat moves in the direction of increasing x .

And $-\lambda\frac{dT}{dx}+\frac{d}{dx}\left(-\lambda\frac{dT}{dx}\right)dx$ is negative when the heat moves in the opposite direction to increasing x .

2.3.2.2. *Balance of conduction terms*

Let us now consider the thin section of unit surface S, of thickness dx, limited by two planes located respectively in x and $x+dx$.

We will set up the **balance of heat entering and exiting the section by conduction.**

When we set up this balance, we do not know[1] what the direction of the variation of $T(x)$ is.

Ox is orientated from left to right, $x<x+dx$

On the left-hand plane, if the heat goes from left to right, it is entering; *entering heat must be counted positively in the balance.*

1 Since we are in the process of establishing the differential equation that will allow the function T(x) itself to be determined!

Therefore, this term will be written $-\lambda \dfrac{dT}{dx}$ per m² because it is positive if the heat moves in the direction of x (from left to right)

Through the plane $x + dx$, the entering heat must also be counted positively in the balance.

Yet $-\lambda \dfrac{dT}{dx} + \dfrac{d}{dx}\left(-\lambda \dfrac{dT}{dx}\right) dx$ is **positive** *when the heat moves in the* **direction of** x, *in other words if* **the heat is exiting** *the section.*

Therefore, this is **the opposite term,** which will intervene in the **balance,** in other words

$$\lambda \frac{dT}{dx} + \frac{d}{dx}\left(\lambda \frac{dT}{dx}\right) dx \qquad [2.21]$$

REMARK.– The manipulation of the signs in front of the terms can surprise or worry certain readers. It is fundamental to understand what is happening.

It must be understood that the negative sign in the expression $-\lambda \dfrac{dT}{dx}$ indicates that the heat moves *physically* from hot to cold and that the flow is positive if it moves in the direction of Ox, whereas the sign used in writing a balance indicates whether the heat *enters* into the volume (necessarily a positive term) or *exits* the volume (necessarily a negative term). Readers are invited to take the time that is required to understand this correctly. This will be of great benefit.

The overall heat balance in the small volume will therefore be:

$$-\lambda \frac{dT}{dx} + \lambda \frac{dT}{dx} + \frac{d}{dx}\left(\lambda \frac{dT}{dx}\right) dx = \frac{d}{dx}\left(\lambda \frac{dT}{dx}\right) dx \qquad [2.22]$$

2.3.2.3. *Variation of the temperature with time*

At this stage, we note that thermal conductibility is associated with the first derivative. The hypothesis of invariable conductibility has not been made.

The term $\frac{d}{dx}\left(\lambda \frac{dT}{dx}\right)dx$ is positive if more heat enters than exits. The small volume heats up.

The term $\frac{d}{dx}\left(\lambda \frac{dT}{dx}\right)dx$ is negative if more heat enters than exits. The small volume cools down.

$\frac{d}{dx}\left(\lambda \frac{dT}{dx}\right)dx$ is exchanged per unit of time. During the time dt, the quantity of heat exchanged with the outside is $\frac{d}{dx}\left(\lambda \frac{dT}{dx}\right)dx\,dt$. It therefore modifies the temperature of the small volume of $d\,T$.

By simply applying the principles of calorimetry, we can write that a change of temperature $d\,T$ of the small volume is due to an exchange of heat $d\,Q$ equal to:

$$d\,Q = \rho\, c\, S\, dx\, d\,T \qquad [2.23]$$

Obviously,

$$d\,Q = \frac{d}{dx}\left(\lambda \frac{dT}{dx}\right)dx\,dt \qquad [2.24]$$

and we have finished our balance:

$$d\,Q = \rho\, c\, S\, dx\, d\,T = \frac{d}{dx}\left(\lambda \frac{dT}{dx}\right)dx\,dt \qquad [2.25a]$$

and we should recall:

$$S = 1 \qquad [2.25b]$$

This gives the single-dimensional expression for the heat equation:

$$\rho c \frac{\partial T}{\partial x} = \frac{\partial}{\partial x}\left(\lambda \frac{\partial T}{\partial x} \right) \qquad [2.26]$$

Let us note that we are now using ∂, because $T = T(x,t)$ and the derivatives are partial.

Two specific cases will intervene in our applications:

a) often the thermal conductibility will be a constant:

$$\frac{\partial T}{\partial x} = a \frac{\partial^2 T}{\partial x^2} \qquad [2.27]$$

This is where thermal diffusivity will come in:

$$a = \frac{\lambda}{\rho c} \qquad [2.28]$$

b) another important case is stationary thermal phenomena. As much heat enters the volume as exits it. Conduction is known as stationary, and the equation becomes:

if the thermal conductibility remains variable:

$$\frac{d}{dx}\left(\lambda \frac{dT}{dx} \right) = 0 \qquad [2.29]$$

We note that $T = T(x)$ and that we have therefore moved to the straight "*d*" that replaces the "∂", in particular partial derivatives.

and for a constant thermal conductibility:

$$\frac{\partial T}{\partial x} = a \frac{\partial^2 T}{\partial x^2} \qquad [2.30]$$

This equation is a specific case of the complete equation as it is established in the Appendix:

$$\frac{\partial T}{\partial t} = \frac{\lambda}{\rho c} div\, gr\vec{a}d\, T \qquad [2.31a]$$

or, for constant physical properties:

$$\frac{\partial T}{\partial t} = a\ \Delta T \qquad [2.31b]$$

2.3.3. *Case of the axisymmetric system*

Here, we give the forms of the heat equation in axisymmetric coordinates in a mono-dimensional case. To lighten the text, we have included details of the calculation in the Appendix, which is based on an approach that is analogous to the previous one.

In this case, the axis becomes an *Oz* axis and the heat (therefore the variation in temperature) follows the vector radius *r*; **positive orientation of each radius from the cylinder axis to the outside of the cylinder.** The same reasoning given previously leads to the following conclusions.

In other words, a single-dimensional field. The temperature depends only on *r*, the distance of a point from the axis. **Isotherms are therefore cylinders with radius** *r*: $T = T(r)$

We no longer think on the basis of a m² of surface area but instead **in terms of a length of 1 m of cylinder.**

$$\rho c \frac{\partial T}{\partial t} = \frac{1}{r} \frac{\partial}{\partial r} \left(\lambda r \frac{\partial T}{\partial r} \right) \qquad [2.32]$$

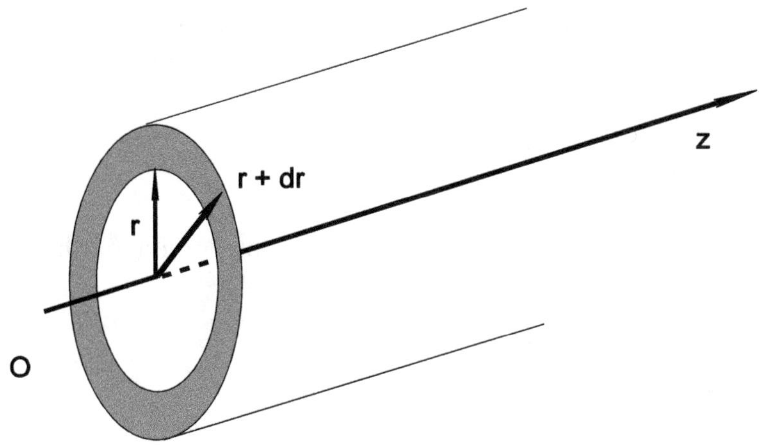

Figure 2.5. *Establishment of the single-dimensional heat equation Axisymmetric geometry.*

With the two specific cases:

– constant physical properties:

$$\frac{\partial T}{\partial t} = \frac{a}{r}\frac{\partial}{\partial r}\left(r\frac{\partial T}{\partial r} \right) \quad [2.33]$$

– stationary conduction:

$$\frac{d}{dr}\left(r\frac{dT}{dr} \right) = 0 \quad [2.34]$$

We note that r is maintained under the first derivative, which is characteristic of equations of cylindrical symmetry.

2.3.4. *Case of the spherical system*

As previously seen, here, we will provide the forms of the heat equation in an axisymmetric coordinate system in a mono-dimensional case. To lighten the text, we have included the details of the calculation in the Appendix, based on the same approach as the one used above.

$$\rho c \frac{\partial T}{\partial t} = \frac{1}{r^2} \frac{\partial}{\partial r} \left(\lambda r^2 \frac{\partial T}{\partial r} \right) \quad [2.35]$$

Here, r is the distance from a point at the origin of the spherical coordinate system.

With the two specific cases:

– constant physical properties:

$$\frac{\partial T}{\partial t} = \frac{a}{r^2} \frac{\partial}{\partial r} \left(r^2 \frac{\partial T}{\partial r} \right) \quad [2.36]$$

– stationary conduction:

$$\frac{d}{dr} \left(r^2 \frac{dT}{dr} \right) = 0 \quad [2.37]$$

We note that r^2 is maintained under the first derivative, which is characteristic of equations with spherical symmetry.

2.4. Resolution of a problem

Heat conduction leads to a very wide range of problems, in terms of their type, structure and difficulty. The problems can be mono-dimensional; the temperature then depends on a single coordinate. In Cartesian coordinates, we have a plane model. In cylindrical coordinates, we have a problem with axial symmetry. These two types will often be encountered in the examples that readers will need to process in this book.

In the general case, the problem is bi- or tri-dimensional.

The problems can also be dependent on time, in which case, they are non-stationary. It is often the case that temperature is independent of time, we then have a stationary problem.

We can see that there will be significant variation in the difficulty of resolving these problems, from a mono-dimensional stationary problem, where $T = T(x)$ or $T = T(r)$ to a three-dimensional non-stationary problem, where, for example, $T = T(x,y,z,t)$ or $T = T(r,\theta,z,t)$. In the first case, we will have analytical solutions that are part of the fundamental culture of a thermal scientist. In the second case, a numerical approach is often required.

Regardless of its type or difficulty, a thermal conduction problem comes down to two things:

A space is defined, in general, in a "region" D defined by a closed surface area S. Through this surface S, heat can be exchanged with space outside the domain. Here, several modes of heat transfer can play a role (conduction, convection, radiation).

The objective of a conduction problem is more often to find the heat flows exchanged between the region and the space outside of it. For this purpose, the temperature distribution near the interface must be known. Beyond this, for example, for reasons of resistance of materials, we are seeking the internal temperature distribution of the region D.

In all cases, and no matter how sophisticated the problem is, there will be two stages in resolving the problem:

1) determination of the temperature domain, generally $T = T(x,y,z,t)$ in the region D;

2) determination of the flows at the interfaces, in general, from local temperature gradients $\vec{grad}\,T\ (x,y,z,t)$ at these interfaces.

Mathematically, the problem will be posed as **the resolution of a heat equation**, in a more or less complete form, associated with **limit conditions**.

The heat equation can be set up by dividing the region into an infinite number of small elements and by writing a heat balance for each of these small elements. We will proceed in this way in the explanations given immediately below. It will then be useful to set up this balance in two simple cases, in the plane problem and in the axisymmetric problem.

A different method, faster and more elegant, consists of using the general methods of differential geometry, which is associated with the previous procedure in a way that is not immediately obvious.

Let us note that this heat equation in the most general case will include time dependence. In addition, if necessary, it should take into account the heterogeneity of the material in the region D.

The conditions at the limits are given by the physical constraints imposed on the region D from the outside. These constraints determine several types of problem, which are most often described as:

– distribution of temperature T_W imposed on the surface S. Mathematically, we have what is known as a **Dirichlet** problem;

– distribution of the thermal flow density φ imposed on the face. This problem can be summarized as imposing the value of the gradient in the material of D on the surface S. Mathematically, we obtain a **Neumann** problem.

Data concerning relationships between temperature and thermal flow density φ on the surface S.

This is particularly the case when the thermal exchange occurs with the space outside the region by conduction, convection or radiation. In this way, we can obtain various examples of the forms seen, since T_e is a reference temperature (distant temperature, external temperature of another material, atmospheric temperature outside D, etc.):

$$\varphi = \lambda \left. \frac{\partial T}{\partial n} \right|_S = \frac{T_W - T_e}{R_{th}} \qquad [2.38]$$

$$\varphi = \lambda \left. \frac{\partial T}{\partial n} \right|_S = h\left(T_W - T_e\right) \qquad [2.39]$$

$$\varphi = \lambda \left. \frac{\partial T}{\partial n} \right|_S = \varepsilon \sigma \left(T_W^4 - T_e^4\right) \qquad [2.40]$$

or a combination of these different expressions. In the presence of convection associated with radiation, we will have, for example:

$$\varphi = \lambda \left.\frac{\partial T}{\partial n}\right|_S = h\left(T_W - T_e\right) + \varepsilon\sigma\left(T_W^4 - T_e^4\right) \qquad [2.41]$$

where h is the convection coefficient, R_{th} is the thermal resistance and $\varepsilon\sigma$ is the product of the Stefan constant and emissivity.

Coupled transfers must also be envisaged:

$$\varphi = \lambda \left.\frac{\partial T}{\partial n}\right|_S = \frac{T_W - T_e}{R_{th}} + h\left(T_W - T_e\right) + \varepsilon\sigma\left(T_W^4 - T_e^4\right) \qquad [2.42]$$

We will see further examples of this later on. Some will be processed in the section of the book on radiation, in particular, in **Chapter 5**.

2.5. Examples of application

NOTE.– The following examples have been placed at this point in this guide, in line with the logical progression of the book and its presentations.

We have just examined the various ways of writing the heat equation.

The problems in this section are direct applications and solutions of this equation. We show, in particular, the effect of a variation of the thermal conductibility with temperature, the effects of the internal generation of heat, and we examine some problems in spherical symmetry.

However, we no longer comply with a logical progression of explanations, which would require us to propose increasingly difficult exercises to readers. Indeed, the problems in this chapter can appear, for a reader with little experience of the field of differential equations, more difficult than the exercises in Chapter **3**, for example.

In this case, readers can go on without prejudice to the exercises in the next few chapters, and when they have acquired a little more agility in resolving the other questions, and have finished reading, they can return to the examples in this chapter.

EXAMPLE 2.1.– Face with variable conductivity

A plane face with thickness e is made up of a homogeneous material whose thermal conductivity can be represented by:

$$\lambda = \lambda_0 (1 + \alpha T) \qquad [2.43]$$

where λ_0 is the thermal conductivity at $T = 0\,°C$. T is therefore expressed in Celsius.

The faces are subject to temperatures T_1 and T_2.

1) Write out the equation which will allow us to determine the distribution of temperature $T = T(x)$ on the face.

2) How does the temperature vary as a function of x? Determine the flow density φ which traverses the face. Deduce from this, the thermal resistance R_{th} of this face.

3) Is the flow lower or higher than what would have been calculated with $\lambda = \lambda_0$?

We will use the following data:

$T_1 = 35\,°C$; $T_2 = 20°C$; $\alpha = 5.10^{-3}\,C^{-1}$; $\lambda_0 = 0.03\,kcal.hr^{-1}.C^{-1}$; $e = 20\,cm$;

SOLUTION.–

1) To solve this problem, we must solve the mono-dimensional heat equation in the stationary regime, written taking into account the variation of the thermal conductibility with temperature.

This equation is written as:

$$\frac{d}{dx}\lambda\frac{d}{dx}T = 0 \qquad [2.44]$$

In this equation:

$$\lambda = \lambda(T) = \lambda_0\ (1 + \alpha T) \qquad [2.45]$$

and $T = T(x)$ is the unknown.

This second-order linear equation requires two limit conditions, which are given by temperatures on the two sides of the face. Choosing the origin on the left-hand face:

$$x = 0 \quad ; \quad T = T_1 \qquad [2.46a]$$

$$x = e \quad ; \quad T = T_2 \qquad [2.46b]$$

2) The differential equation can be re-written as:

$$\frac{d}{dx}\lambda_0\ (1 + \alpha T)\frac{d}{dx}T = 0 \qquad [2.47]$$

The constant λ_0 can be eliminated.

The equation will be solved by integrating twice by x:

$$\frac{d}{dx}(1 + \alpha T)\frac{d}{dx}T = 0 \qquad [2.48]$$

The first integration leads to a first constant C_1:

$$(1 + \alpha T)\frac{d}{dx}T = C_1 \qquad [2.49]$$

We note that $(1 + \alpha T)\frac{d}{dx}T$ can be written as the derivative of a function:

$$(1 + \alpha T)\frac{d}{dx}T = \frac{dT}{dx} + \alpha T\frac{dT}{dx} = \frac{dT}{dx} + \frac{d}{dx}\alpha\frac{T^2}{2} = \frac{d}{dx}\left(T + \alpha\frac{T^2}{2}\right) \qquad [2.50]$$

The equation can be re-written as:

$$d\left(T+\alpha\frac{T^2}{2}\right)=C_1\,dx \qquad [2.51]$$

which is integrated into:

$$\left(T+\alpha\frac{T^2}{2}\right)=C_1\ x+C_2 \qquad [2.52]$$

And determined by the limit conditions:

$$x=0 \quad ; \quad T=T_1 \text{ therefore } \left(T_1+\alpha\frac{T_1^2}{2}\right)=C_1\ *0+C_2 \qquad [2.53]$$

$$x=e \quad ; \quad T=T_2 \text{ therefore } \left(T_2+\alpha\frac{T_2^2}{2}\right)=C_1\ *e+C_2 \qquad [2.54]$$

It results that:

$$C_2=\left(T_1+\alpha\frac{T_1^2}{2}\right) \qquad [2.55]$$

$$C_1\ =\frac{1}{e}\left[\left(T_2+\alpha\frac{T_2^2}{2}\right)-\left(T_1+\alpha\frac{T_1^2}{2}\right)\right] \qquad [2.56]$$

The numerical values of C_1 and C_2 are:

$$C_2\ =\ 38.06 \qquad [2.57a]$$

$$C_1\ =-85.3 \qquad [2.57b]$$

To calculate the flow densities, it is not necessary to express $T(x)$ explicitly.

In fact, the first integration has given us:

$$(1+\alpha T)\frac{d}{dx}T = C_1 \qquad [2.58]$$

which is the flow density at the nearest λ_0.

Therefore,

$$\varphi = \lambda_0 (1+\alpha T)\frac{d}{dx}T = C_1\,\lambda_0 \qquad [2.59]$$

$$\varphi = 85.3 * 0.0348 = 2.97\,W.m^{-2} \qquad [2.60]$$

The thermal resistance of the wall can be deduced by:

$$R_{th} = \frac{T_1 - T_2}{\varphi} \qquad [2.61]$$

$$R_{th} = \frac{15}{2.97} = 5.05\,m^2.K.W^{-1} \qquad [2.62]$$

3) To compare the real flows to those calculated with $\lambda = \lambda_0$, two methods can be used.

a) Compare the flow density found in 2 directly to the flow density φ_0, which traversed a wall of thermal conductibility λ_0.

$$\varphi_0 = \lambda_0 \frac{T_1 - T_2}{e} \qquad [2.63]$$

In an SI coordinate system:

$$\lambda_0 = 0.03\ kcal.hr^{-1}.K^{-1} = \frac{0.03 * 4180}{3600} = 0.0348\,W.m^{-1}.K^{-1} \qquad [2.64]$$

$$\varphi_0 = 2.61\,W$$

The real flow is higher than this value. The real wall is less insulating than the homogeneous wall.

We note that we can write

$$\varphi_0 = \frac{T_1 - T_2}{R_{th0}} \text{ with } R_{th0} = \frac{e}{\lambda_0} \quad [2.65]$$

$$\varphi_0 = 0.0348 \frac{15}{0.2} = 2.61 W.m^{-2} \quad [2.66]$$

b) Compare the thermal resistance found in 2 to R_{th0}, thermal resistance of a wall of thermal conductibility λ_0

$$R_{th0} = \frac{e_2}{\lambda_0} = \frac{0.2}{0.0348} = 5.75\, m^2.K.W^{-1} \quad [2.67]$$

The resistance of the real wall $R_{th} = 5.05\, m^2.K.W^{-1}$ comes out lower than this value. We confirm that the real wall is less insulating than the homogeneous wall.

EXAMPLE 2.2.– Problems with generation of internal heat. Evacuation of heat in a bar of uranium

A bar of uranium has a diameter $D = 29\, mm$.

The nuclear reactions that take place here produce a volumetric power q expressed in $qW.m^{-3}$.

The thermal conductivity of uranium is $\lambda = 27 W.m^{-1}.K^{-1}$.

1) Determine the distribution in the bar under the stationary regime. At the external surface, the temperature is $T_e = 200\,°C$. What is the maximum temperature $T_{\max}$ obtained? Where do we observe this?

2) Uranium melts at $T_f = 1132\,°C$. Determine the maximum volumetric power $q_{\max}$ that can be extracted from the bar if we do not want to make the uranium melt in the center of the power plant, which is always best to avoid?

SOLUTION.–

Here, we must resolve the heat equation, written in cylindrical coordinates with a volumetric production term.

The problem has rotational symmetry; therefore, $T = T(r)$

$$\frac{\lambda}{r}\frac{d}{dr}\left(r\frac{dT}{dr}\right) + q = 0 \quad [2.68a]$$

We know one limit condition.

In principle, this is insufficient, but we will see another piece of information come to light later on.

$$R = \frac{D}{2} = 14.5\,mm \quad [2.68b]$$

$$r = R \quad ; \quad T = T_e \quad [2.69]$$

The equation can be solved by integrating twice:

$$\frac{d}{dr}\left(r\frac{dT}{dr}\right) = \frac{-q\,r}{\lambda} \quad [2.70]$$

$$r\frac{dT}{dr} = \frac{-q\,r^2}{2\lambda} + C_1 \quad [2.71]$$

$$\frac{dT}{dr} = \frac{-q\,r}{2\lambda} + \frac{C_1}{r} \quad [2.72]$$

$$T(r) = \frac{-q\,r^2}{4\lambda} + C_1\,Ln\,r + C_2 \quad [2.73]$$

The temperature must remain finite everywhere in the tube, which is not the case in the center, unless we use

$$C_1 = 0 \qquad [2.74]$$

C_2 is then easily deduced from the external temperature of the tube:

$$T_e = \frac{-q\,R^2}{4\lambda} + C_2 \qquad [2.75]$$

$$C_2 = T_e + \frac{q\,R^2}{4\lambda} \qquad [2.76]$$

Finally, we can put the temperature in the form:

$$T(r) - T_e = \frac{q\,R^2}{4\lambda}\left[1 - \left(\frac{r}{R}\right)^2\right] \qquad [2.77]$$

We obtain a parabolic temperature profile. The maximum temperature $T_{\max}$ is obtained for $r = 0$, in other words in the center of the tube. The maximum temperature reached is:

$$T_{\max} = T_e + \frac{q\,R^2}{4\lambda} \qquad [2.78]$$

2) If we do not want the temperature of the bar to exceed $1132°\,C$, we find the maximum value of the volumetric power to be:

$$T_F = T_e + \frac{q_{\max}\,R^2}{4\lambda} \qquad [2.79]$$

In other words

$$q_{\max} = \frac{4\lambda(T_F - T_e)}{R^2} \qquad [2.80]$$

$$q_{\max} = \frac{4*27*(1132-200)}{\left(14.5.10^{-3}\right)^2} = 4.79.10^8\ W.m^{-3} \quad [2.81]$$

EXAMPLE 2.3.– Maximum dissipation of an electrical resistance

An electrical resistance takes the form of a metal tube with diameter $d = 2\,cm$ surrounded by an insulating duct of thickness $e = 5\,mm$. The thermal conductibility of this duct is $\lambda = 0.06\,m^{-1}\,K^{-1}$.

In other words, $h_i = 5\,W\,m^{-2}\,K^{-1}$ the convection coefficient outside the system.

We do not want the metal part of the resistance to exceed $T_{\max} = 150\,°C$. The atmospheric temperature is $T_e = 20\,°C$.

What power can dissipate this resistance per meter?

SOLUTION.–

Let us calculate the resultant thermal resistance of the insulation and convection:

– the internal radius of the insulator is $r_i = \frac{d}{2}$

– the external radius is $r_e = r_i + e$

$$R_{th} = \frac{Ln\left(\frac{r_i + e}{r_i}\right)}{2\,\pi\,\lambda} + \frac{1}{2\,\pi\,(r_i + e)\,h} = \frac{Ln\left(\frac{0.01+0.005}{0.01}\right)}{2\,\pi\,0.06} + \frac{1}{2\,\pi * 0.015 * 5} \quad [2.82]$$

$$R_{th} = 1.075 + 2.12 = 3.197\ m\,K\,W^{-1} \quad [2.83]$$

In other words, the flow per meter that leads to a temperature $T_{\max} = 150\,°C$:

$$\Phi = \frac{\Delta T}{R_{th}} = \frac{150-20}{3.197} = 40.66\,W \quad [2.84]$$

EXAMPLE 2.4.– Temperature in steel wire, which an electric current is passed through.

We consider a steel wire with a diameter $d = 2\,mm$. Its thermal conductibility is $\lambda = 54\,m^{-1}\,K^{-1}$. An electric current is passed through it, which dissipates a power per ml of wire by the Joule effect $P = 100\,W\,m^{-1}$. This power is dissipated homogeneously in the mass of the wire. In other words, q, the resulting volumetric power, is expressed in $W\,m^{-3}$.

The external temperature of the wire is maintained at $T_e = 60\,°C$.

1) What is the value of *q*?

2) Write the balance equation in a small circular volume of unit length and thickness $d\,r$, included between cylinders of radius r and $r + d\,r$.

3) Solve this equation and give the expression for $T(r)$.

4) What is the maximum value of temperature $T_{\max}$ in the wire. Where is it located?

SOLUTION.–

1) The volume of a meter of wire is:

$$V_{OL} = 1 * \frac{\pi\, d^2}{4} = \frac{\pi\, 4.10^{-6}}{4} = 3.14.10^{-6}\ m^3 \qquad [2.85]$$

and is therefore equal to:

$$q = \frac{P}{V_{OL}} = \frac{100}{3.14.10^{-6}} = 3.18.10^{6}\ W\,m^{-3} \qquad [2.86]$$

2) We write a *classic heat balance for* $L = 1\,m$ *of wire*

$$-2\pi\,\lambda\frac{dT}{dr} - \left[-2\pi\,\lambda\frac{dT}{dr} + \frac{d}{dr}\left(-2\pi\,\lambda\frac{dT}{dr}\right)\right] + 2\pi\, r\, dr\, q = 0 \qquad [2.87]$$

Dividing the two terms by $2\pi\lambda$

$$\frac{1}{r}\frac{d}{dr}r\frac{dT}{dr}+\frac{q}{\lambda}=0 \qquad [2.88]$$

$$\frac{1}{r}\frac{d}{dr}r\frac{dT}{dr}+\frac{q}{\lambda}=0 \qquad [2.89]$$

associated with a limit condition that we have imposed at the surface of the wire:

$$r=R \quad ; \quad T=T_W \qquad [2.90]$$

3) Resolution of this equation is of a type that is often found physically:

$$\frac{d}{dr}r\frac{dT}{dr}=-\frac{qr}{\lambda} \qquad [2.91]$$

Integrating the two terms

$$r\frac{dT}{dr}=-\frac{qr^2}{2\lambda}+C_1 \qquad [2.92]$$

$$\frac{dT}{dr}=-\frac{qr}{2\lambda}+\frac{C_1}{r} \qquad [2.93]$$

$$T=-\frac{qr^2}{4\lambda}+C_1\ Ln\, r+C_2 \qquad [2.94]$$

T must remain finite at the center of the wire, therefore: $C_1=0$

At the surface,

$r=R$, we find $T=T_w$ [2.95]

Therefore:

$$T_W=-\frac{qR^2}{4\lambda}+C_2 \qquad [2.96]$$

Finally:

$$T - T_w = \frac{qR^2}{4\lambda}\left[1 - \left(\frac{r}{R}\right)^2\right] \qquad [2.97]$$

4) We see that the temperature will be a maximum for $x = 0$, in other words, the center of the wire:

For $r = 0$

$$T_{max} - T_w = \frac{qR^2}{4\lambda} \qquad [2.98]$$

$$T_{max} - T_w = \frac{qR^2}{4\lambda} = \frac{3.18.10^6 * 10^{-6}}{4 * 54} = 1.47.10^{-2}\ °C \qquad [2.99]$$

Hence, the maximum temperature at the center:

$$T_{max} - T_w = 60 + 1.47.10^{-2} = 61.47°C \qquad [2.100]$$

2.5.1. *Problems involving spherical symmetry*

EXAMPLE 2.5.– Thermal resistance of two concentric spheres

We consider two concentric spheres with radius r_1 and r_2, maintained at constant temperatures T_1 and T_2.

The solid environment that separates them is homogeneous, isotropic and presents a thermal conductivity λ that is independent of the temperature.

1) Calculate the thermal power dissipated in the stationary regime.

2) What is the thermal resistance of the system?

SOLUTION.–

1) The thermal power sought is equal to the incoming flow in the sphere of radius r_1. This flow is equal to the flow traversing through each intermediate sphere between r_1 and r_2. It is also equal to the flow that traverses the sphere of radius r_2.

For a value of r between r_1 and r_2, the general expression of this flow is:

$$\Phi = 4\pi \lambda r^2 \frac{dT}{dr} \quad [2.101]$$

And in particular, at the interfaces of the region contained between the two spheres of radius r_1 and r_2:

$$\Phi = -4\pi \lambda r^2 \frac{dT}{dr}\bigg|_{r=r_1} = -4\pi \lambda r^2 \frac{dT}{dr}\bigg|_{r=r_2} \quad [2.102]$$

The distribution of $T(r)$ must therefore be determined in advance.

It must be deduced by writing a thermal balance that will lead to a differential equation.

By writing that the incoming and outgoing flows of a thin spherical section of thickness dr are equal, we obtain:

$$-\lambda \frac{dT}{dr} 4\pi r^2 + \lambda \frac{dT}{dr} 4\pi r^2 + \frac{d}{dx}\left(\lambda 4\pi r^2 \frac{dT}{dr} \right) dr = \frac{d}{dr}\left(\lambda 4\pi r^2 \frac{dT}{dr} \right) = 0 \quad [2.103]$$

In other words, for a homogeneous body of constant thermal conductibility:

$$\frac{d}{dr}\left(r^2 \frac{dT}{dr} \right) = 0 \quad [2.104]$$

With the limit conditions

$$r = r_1 \quad ; \quad T = T_1 \quad [2.105a]$$

$$r = r_2 \quad ; \quad T = T_2 \quad [2.105b]$$

We note the term included below the derivative symbol. With this written expression, we take into account the variation of the spherical surface traversed by the flow when r varies from r to $r+dr$.

Resolution of this equation is obtained by integrating twice

$$\frac{d}{dr}\left(r^2\frac{dT}{dr}\right)=0 \qquad [2.106]$$

$$r^2\frac{dT}{dr}=C_1 \qquad [2.107]$$

$$\frac{dT}{dr}=\frac{C_1}{r^2} \qquad [2.108]$$

$$T(r)=-\frac{C_1}{r}+C_2 \qquad [2.109]$$

The constants will be determined by the limit conditions

$$r=r_1 \quad ; \quad T=T_1 \quad ; \quad T_1=-\frac{C_1}{r_1}+C_2 \qquad [2.110]$$

$$r=r_2 \quad ; \quad T=T_2 \quad ; \quad T_2=-\frac{C_1}{r_2}+C_2 \qquad [2.111]$$

Subtracting the two expressions:

$$T_1-T_2 = -C_1\left(\frac{1}{r_1}-\frac{1}{r_2}\right) \qquad [2.112a]$$

$$C_1=-\frac{T_1-T_2}{\left(\frac{1}{r_1}-\frac{1}{r_2}\right)} \qquad [2.112b]$$

We note that if $T_1>T_2$, $T_1-T_2>0$ and C_1 is negative.

C_2 can then be written in two different ways:

$$C_2 = T_1 + \frac{C_1}{r_1} \quad [2.113]$$

$$C_2 = T_2 + \frac{C_1}{r_2} \quad [2.114]$$

The flow in r_1 or in r_2 is written:

$$\Phi = -4\pi\lambda r^2 \frac{dT}{dr}\bigg|_{r=r_1} = -4\pi\lambda r^2 \frac{dT}{dr}\bigg|_{r=r_1} \quad [2.115]$$

$$\Phi = -4\pi\lambda r^2 \frac{dT}{dr}\bigg|_{r=r_1} = -4\pi\lambda r^2 \frac{d}{dr}\left(-\frac{C_1}{r} + C_2\right)\bigg|_{r=r_2}$$

$$= -4\pi\lambda r^2 \frac{C_1}{r^2}\bigg|_{r=r_1} = -4\pi\lambda r^2 C_1 \quad [2.116]$$

$$\Phi = -4\pi\lambda r^2 \frac{dT}{dr}\bigg|_{r=r_2} = -4\pi\lambda r^2 \frac{C_1}{r^2}\bigg|_{r=r_2} = -4\pi\lambda r^2 C_1 \quad [2.117]$$

We verify that the two values are indeed identical

$$\Phi = -4\pi\lambda r^2 C_1 = 4\pi\lambda r^2 \frac{T_1 - T_2}{\left(\dfrac{1}{r_1} - \dfrac{1}{r_2}\right)} \quad [2.118]$$

2) By definition of the thermal resistance R_{th} per m²:

$$\Phi = 4\pi r^2 \frac{T_1 - T_2}{R_{th}} \quad [2.119]$$

By identification:

$$R_{th} = \frac{(r_2 - r_1)}{\lambda r_1 r_2} \qquad [2.120]$$

EXAMPLE 2.6.– Conduction in a sphere: Whittaker model

Whittaker proposes the following general expression for the Nusselt number, which allows the thermal transfer between a full isothermal sphere of radius R and diameter D, at a constant temperature T_w and an atmosphere at temperature T_e :

$$N_u = 2 + \left[0.4\, R_e^{\frac{1}{2}} + 0.06\, R_e^{\frac{2}{3}} \right] P_r^{0.4} \left(\frac{\mu_e}{\mu_w} \right) \qquad [2.121]$$

where R_e is the Reynolds number based on the diameter of the sphere and μ is the dynamic viscosity.

This equation will be valid for $3.5 < R_e < 80\,000$ and $0.7 < P_r < 380$

Justify term 2 in this equation.

SOLUTION.–

The second term gives the value of the Nusselt number when the Reynolds numbers are zero. This corresponds to the absence of flow around the sphere. Therefore, there is no longer any convective transport and term 2 corresponds to a purely conductive transfer, which in all cases coexists with convection when the latter is present.

REMARK.– We will see in the following that this transfer by conduction, in this spherical symmetry system, will be stationary. This would not be the case with a plane face; a non-stationary transfer would then be necessary.

We calculate the exchange, by conduction, between a full sphere of radius R at constant temperature T_w and an atmosphere at temperature T_e .

We are in a single-dimensional stationary regime: $T = T(r)$. The isotherms are spheres with radius r.

We will now write the balance for the space found between two spheres of radii r and $r + dr$:

$$\lambda \frac{d}{dr} \pi r^2 \frac{dT}{dr} = 0 \qquad [2.122]$$

which is reduced to the differential equation:

$$\lambda \frac{d}{dr} \pi r^2 \frac{dT}{dr} = 0 \qquad [2.123]$$

with the limit conditions

$$r = R \quad ; \quad T = T_W \qquad [2.124a]$$

$$r \to \infty \quad ; \quad T \to T_e \qquad [2.124b]$$

We will consider the new function:

$$\theta(r) = T - T_e \qquad [2.125]$$

Defining:

$$\theta_0 = T_w - T_e \qquad [2.126]$$

The system to be resolved is transformed into:

$$\frac{d}{dr} r^2 \frac{dT}{dr} = 0 \qquad [2.127]$$

with the conditions:

$$r = R \quad ; \quad \theta = \theta_0 \qquad [2.128a]$$

$$r \to \infty \quad ; \quad \theta \to 0 \qquad [2.128b]$$

Integrating the equation twice, two constants C_1 and C_2 appear:

$$r^2 \frac{d\theta}{dr} = C_1 \quad [2.129]$$

$$\frac{d\theta}{dr} = \frac{C_1}{r^2} \quad [2.130]$$

$$\theta(r) = \frac{-C_1}{r} + C_2 \quad [2.131]$$

Applying the limit conditions:

$$r = R \quad ; \quad \theta_0 = \frac{-C_1}{R} + C_2 \quad [2.132]$$

$$r \to \infty \quad ; \quad 0 = C_2 \quad [2.133]$$

which gives:

$$C_1 = -R\theta_0 \quad [2.134]$$

$$\theta(r) = \frac{R\theta_0}{r} \quad [2.135]$$

Therefore, the thermal flow density for the sphere:

$$\varphi = -\lambda \left.\frac{dT}{dr}\right|_{r=R} = -\lambda \left.\frac{d\theta}{dr}\right|_{r=R} = -\lambda \left.\frac{-R\theta_0}{r^2}\right|_{r=R} = \lambda \frac{R\theta_0}{R^2} = \frac{\lambda\theta}{R}{}_0 \quad [2.136]$$

which gives a convection coefficient

$$h = \frac{\varphi}{T_w - T_a} = \frac{\varphi}{\theta_0} = \frac{\lambda}{R} = \frac{2\lambda}{D} \quad [2.137]$$

We deduce the classic expression for the Nusselt number from this:

$$N_u = \frac{hD}{\lambda} \qquad [2.138]$$

$$N_u = 2 \qquad [2.139]$$

This Nusselt number is only valid for an exclusively conductive transfer. In general, convective terms are added to it, which take the form of polynomials with powers of the Reynolds numbers of the sphere.

EXAMPLE 2.7.– Evaporation of a drop: Godsave model

A spherical drop of liquid, radius R and diameter D, isotherm at a temperature T_w, is immobile in an atmosphere at temperature T_e, at which it vaporizes.

In other words, ρ_L is the density of the liquid, L_V is the heat of vaporization of the mass and λ is the thermal conductibility of air. $T_e > T_L > T_a$, where T_V is the vaporization temperature of the liquid (saturating vapor temperature at atmospheric pressure).

We suppose that the surface of the drop remains equal to T_V throughout the vaporization.

1) Considering that all of the heat brought to the drop is used to vaporize the liquid at the surface, write the variation in time of the mass of the drop.

2) Deduce from this the differential equation that the diameter of the drop obeys.

3) Show that the reduction of the diameter of the drop is linear:

$$D(t) = D_0(t) - Kt$$

where D_0 is the initial diameter of the drop and K is a constant for which the expression will be given.

Deduce the vaporization time τ_V for the drop.

We will use a quasi-stationary calculation, supposing that, even for slow changes to the diameter, the Nusselt number taken from the Whittaker equation remains valid:

$$N_u = 2 + \left[0.4\, R_e^{\frac{1}{2}} + 0.06\, R_e^{\frac{2}{3}} \right] P_r^{0.4} \left(\frac{\mu_e}{\mu_w} \right) \qquad [2.140]$$

This equation would be valid for $3.5 < R_e < 80\,000$ and $0,7 < P_r < 380$

SOLUTION.–

1) The thermal flow Φ_W provided to the drop will, at instant t, have the form:

$$\Phi_W = S(t)\, h(T_a - T_V) \qquad [2.141]$$

where $S(t)$ is the surface area of the surface of the drop:

$$S(t) = 4\pi R^2 = 2\pi D^2 \qquad [2.142]$$

This flow, used to vaporize the liquid, allows a mass $\dot{m}_v$ to be transformed, per unit of time, into vapor:

$$\dot{m}_v = \rho_L \Phi_w\, L_V \qquad [2.143]$$

$$\dot{m}_v = \rho_L\, L_V\, 2\pi D^2 h(T_a - T_V) \qquad [2.144]$$

The instantaneous mass of the drop is $m(t)$:

$$m(t) = \frac{4}{3}\pi R^3 = \pi \frac{D^3}{6} \qquad [2.145]$$

If the drop is immobile, the Whittaker equation is reduced to

$$N_u = 2 \qquad [2.146]$$

This result, which can be deduced directly (see previous exercise), results from a written theory in a stationary regime. When $R(t)$ varies slowly with temperature, we suppose that this Nusselt value remains valid.

Here, we will continue the development, maintaining the notation N_u, even if we know the value for an immobile fluid. The explanation for this will be given at the end of the solution.

h will be deduced from the Nusselt definition:

$$N_u = \frac{h D}{\lambda} \quad [2.147]$$

$$h = \frac{N_u \lambda}{D} \quad [2.148]$$

Finally:

$$\dot{m}_v = \rho_L L_V 2\pi D^2 \frac{N_u \lambda}{D}(T_a - T_V) = \rho_L L_V 2\pi D N_u \lambda (T_a - T_V) \quad [2.149]$$

The balance for the mass of the drop is therefore written as:

$$\dot{m}_v = -\frac{dm}{dt} = -\rho_L \frac{d}{dt}\frac{\pi D^3}{6} \quad [2.150]$$

$$\dot{m}_v = -\rho_L \frac{D^2}{2}\frac{dD}{dt} \quad [2.151]$$

2) By deduction, combining the two expressions for $\dot{m}_v$:

$$\dot{m}_v = \rho_L L_V 2\pi D N_u \lambda (T_a - T_V) = -\rho_L \frac{D^2}{2}\frac{dD}{dt} \quad [2.152]$$

$$D\frac{dD}{dt} = -L_V 4\pi N_u \lambda (T_a - T_V) \quad [2.153]$$

That we can re-write, by making the time derivative for D^2 appear:

$$\frac{1}{2}\frac{d\,D^2}{d\,t} = -\,L_V\,4\pi\;N_u\,\lambda\left(T_a - T_V\right) \qquad [2.154]$$

where:

$$\frac{d\,D^2}{d\,t} = -\,L_V\,8\pi\;N_u\,\lambda\left(T_a - T_V\right) \qquad [2.155]$$

In other words:

$$\frac{d\,D^2}{d\,t} = -\,K \qquad [2.156]$$

Writing:

$$K = L_V\,8\pi\;N_u\,\lambda\left(T_a - T_V\right)$$

3) The linear equation can easily be resolved into

$$D^2 = -\,Kt + C \qquad [2.157]$$

with the initial condition:

$$t = 0\;;\;D(0) = D_0 \qquad [2.158]$$

Hence the law:

$$D^2 = D_0^2 - Kt \qquad [2.159]$$

The vaporization time τ_V is reached when the diameter of the drop is zero:

$$0 = D_0^2 - K\,\tau_V \qquad [2.160]$$

$$\tau_V = \frac{D_0^2}{K} \qquad [2.161]$$

with

$$K = L_V \, 8\pi \, N_u \, \lambda(T_a - T_V) \qquad [2.162]$$

This law is known under the name of the law expressed in D^2. We effectively note that the vaporization time is proportional to the square of the initial diameter and proportional to the temperature differential between atmosphere and saturation point.

When $N_u = 2$, we have:

$$K = 16\pi \, L_V \, \lambda(T_a - T_V) \qquad [2.163]$$

$$\tau_V = \frac{D_0^2}{16\pi \, L_V \, \lambda(T_a - T_V)} \qquad [2.164]$$

REMARK.– We have kept N_u until the end of the calculation because this model can still be applied in the presence of wind blowing across the drop. In this case, use the form proposed by Whittaker:

$$N_u = 2 + \left[0.4 \, R_e^{\frac{1}{2}} + 0.06 \, R_e^{\frac{2}{3}} \right] P_r^{0.4} \left(\frac{\mu_e}{\mu_w} \right) \qquad [2.165]$$

We see that in this way, a vaporization drop continues to satisfy a law for D^2. Convection obviously has the effect of increasing the constant K, therefore reducing the vaporization time. This result can be applied in various practical situations: reactors in chemical engineering, agriculture (spraying of pesticides), injection of petrol or diesel into a cylinder, rocket engines, etc.

3

Conduction in a Stationary Regime

3.1. Thermal resistance

This practical concept is applicable to plane and cylindrical geometries. We will demonstrate that, depending on the composition or geometry of an insulating assembly, several definitions for thermal resistance will be generated, and we must be careful not to confuse them.

3.1.1. *Thermal resistance: plane geometry*

In some practical calculations, we obtain the notion of thermal resistance using the "simplified" model of a plane wall.

It is important, at this point, to define the resistance that we apply more carefully.

In fact, there are two thermal resistances that are very different:

a) the first is defined per unit of area of a wall. We will refer to it as "thermal resistance", which is denoted by R_{th}. We will use this almost systematically in this chapter. We prefer to use this term because for a homogeneous wall, this resistance is also a characteristic of the material it is made of. It is generally related to the calculation of flow density.

b) the second is defined for a given surface of the wall. We will refer to it as "global resistance", which is denoted by R_G.

In general, it is related to the calculation of a flow.

3.1.1.1. *Resistance per unit of surface area (per m²)*

We know that the density of thermal flow φ that passes through a plane wall of thickness e made of a material with thermal conductibility λ, to which we apply a temperature difference $T_1 - T_2$, is given by:

$$\varphi = \lambda \frac{T_1 - T_2}{e} \quad [3.1]$$

If S is the surface area of the wall, the flow Φ that traverses it is then

$$\Phi = S\,\varphi = \lambda\, S \frac{T_1 - T_2}{e} \quad [3.2]$$

We can rewrite the thermal flow density as:

$$\varphi = \frac{T_1 - T_2}{\dfrac{e}{\lambda}} = \frac{T_1 - T_2}{R_{th}} \quad [3.3]$$

We have therefore defined the thermal resistance per unit of surface area, R_{th}, or thermal resistance per m²:

$$R_{th} = \frac{e}{\lambda} \quad [3.4]$$

We note that for a wall made of a unique homogeneous material, this thermal resistance thus defined is a characteristic of the material. Moreover, this characteristic is often used for commercial purposes.

With the same idea in mind, we also define a conductance U:

$$U = \frac{1}{R_{th}} = \frac{\lambda}{e} \quad [3.5]$$

Conductance can also be used as commercial product data. For the sake of simplicity, we will not use this parameter in examples proposed in this book:

$$R_{th} = \frac{e}{\lambda} \text{ is expressed in } m^2 K W^{-1} \quad [3.6a]$$

$$U = \frac{\lambda}{e} \text{ is expressed in } W\,m^{-2}\,K^{-1} \quad [3.6b]$$

3.1.1.2. *Overall resistance*

Here, we start with the expression for a flow Φ that traverses an area S of a plane wall with thickness e that is made of a material with thermal conductibility λ, to which a temperature difference $T_1 - T_2$ is applied, which is given by:

$$\Phi = S\,\varphi = \lambda\,S\,\frac{T_1 - T_2}{e} \quad [3.7]$$

which can be rewritten as:

$$\Phi = \frac{T_1 - T_2}{\dfrac{e}{\lambda S}} = \frac{T_1 - T_2}{R_G} \quad [3.8]$$

We have thus defined the overall thermal resistance, R_G, or the thermal resistance per m²:

$$R_G = \frac{e}{\lambda S} \quad [3.9]$$

We note that this resistance can no longer be characteristic of a material, but depends on the surface area of the wall.

Similarly, we can also define a conductance U_G:

$$U = \frac{1}{R_G} = \frac{\lambda S}{e} \quad [3.10]$$

$$R_{th} = \frac{e}{\lambda} \text{ is expressed in } m^2\,K\,W^{-1} \quad [3.11a]$$

$$U = \frac{\lambda}{e} \text{ is expressed in } W\,m^{-2}\,K^{-1} \quad [3.11b]$$

3.1.1.3. *Practical application of thermal resistance*

The concept of thermal resistance becomes useful when considering practical cases of walls made up of superimposed materials of different types and thicknesses.

We then distinguish between walls made up of superimposed layers and those made up of composite materials.

a) Plane walls of superimposed layers: addition of thermal resistances

In this common case, the wall, for which we want to determine the insulation performance, is made up of layers. We can still reason in terms of density of thermal flow, because this is the same density that passes through all materials.

The notion of thermal resistance per unit of surface area will therefore be given preference. We can therefore obviously define a thermal resistance

$$R_{thi} = \frac{e_i}{\lambda_i} \quad [3.12]$$

for each component of the wall and a "resultant" thermal resistance R_{th} of the wall as a whole.

We will demonstrate this in an example.

EXAMPLE 3.1.– Resultant thermal resistance of a plane wall

What is the resultant thermal resistance of a plane wall composed of three layers of materials with respective thermal conductibilities λ_1, λ_2 and λ_3, and respective thicknesses e_1, e_2 and e_3?

SOLUTION.–

We can set out the thermal resistance (per unit of surface area) for each of the layers:

$$R_{th1} = \frac{e_1}{\lambda_1} \quad [3.13a]$$

$$R_{th2} = \frac{e_2}{\lambda_2} \quad [3.13b]$$

$$R_{th3} = \frac{e_3}{\lambda_3} \quad [3.13c]$$

We will denote the extreme temperatures applied to the two faces of the wall as T_1 and T_4, and the temperatures of the intermediate interfaces between the layers of the wall as T_2 and T_3.

The thermal flow density T_1 can be written for each of the materials as:

$$\varphi = \frac{T_1 - T_2}{R_{th1}} \quad [3.14]$$

$$\varphi = \frac{T_2 - T_3}{R_{th2}} \quad [3.15]$$

$$\varphi = \frac{T_3 - T_4}{R_{th3}} \quad [3.16]$$

We express the differences in temperature and add the three expressions obtained:

$$T_1 - T_2 = \varphi R_{th1} \quad [3.17]$$

$$T_2 - T_3 = \varphi R_{th2} \quad [3.18]$$

$$T_3 - T_4 = \varphi R_{th3} \quad [3.19]$$

$$(T_1 - T_2) + (T_2 - T_3) + (T_3 - T_4) = \varphi(R_{th1} + R_{th2} + R_{th3}) \quad [3.20]$$

$$\varphi = \frac{T_1 - T_4}{R_{th1} + R_{th2} + R_{th3}} = \frac{T_1 - T_4}{R_{th}} \quad [3.21]$$

Finally, we obtain:

$$\varphi = \frac{T_1 - T_4}{R_{th1} + R_{th2} + R_{th3}} = \frac{T_1 - T_4}{R_{th}} \quad [3.22]$$

which leads us to the relationship between the overall thermal resistance R_{th} and the thermal resistances of each wall:

$$R_{th} = R_{th1} + R_{th2} + R_{th3} \quad [3.23]$$

We also note here a thermal analogy with Ohm's law, assimilating the density of thermal flow to an intensity and the temperatures to potentials.

b) Composite wall

In this case, the wall is no longer made up of layers that are simply superimposed: it is heterogeneous. A problem of this kind, tackled from a rigorous point of view, involves resorting to a numerical method. We can find an approximate solution (more or less) to this problem using the notion of overall thermal resistance.

As an example, let us consider the case of a wall made up of a concrete base with an upper part made up of bricks, covered by two layers of insulating material.

EXAMPLE 3.2.– Composite wall

More precisely, a wall of thickness e_2 is made up of a concrete base with thermal conductibility λ_C and a surface area S_1, and a face (of the same thickness e_2) with thermal conductibility λ_B and a surface area S_2. Each of the two faces of the entire wall is covered with a layer of insulation of thermal conductibility λ_I and thickness e_1.

We apply a temperature difference $T_1 - T_2$ between the two external faces of this composite wall. What is the flow passing through the wall?

SOLUTION.–

We divide the composite wall into two plane walls made up of plane layers:

– for the first wall, the concrete assembly and the two layers of insulation;

– for the second wall, the brick wall and its two layers of insulation.

Each of these walls can be treated using an analogous approach to that previously applied for the plane multilayer wall; but this time by writing the expression for the two flows Φ_1 and Φ_2 passing through the concrete and the brick respectively, in terms of overall thermal resistances. This is made possible by the surface area that the three layers have in common.

We can define various overall thermal resistances for each layer of each of these multilayer walls.

For the base wall, with a surface area S_1

concrete:

$$R_G = \frac{e_2}{\lambda_C S_1} \qquad [3.24]$$

each layer of insulation:

$$R_{GI} = \frac{e_1}{\lambda_I S_1} \qquad [3.25]$$

For the brick wall, with a surface area S_2

brick:

$$R_{GB} = \frac{e_2}{\lambda_B S_2} \qquad [3.26]$$

each layer of insulation:

$$R_{GI2} = \frac{e_1}{\lambda_I S_2} \qquad [3.27]$$

For each of the two sections of the composite wall, we therefore find:

$$R_{G1} = R_{GI1} + R_{GC} + R_{GI1} \qquad [3.28]$$

$$R_{GI2} = R_{GI2} + R_{GB} + R_{GI2} \qquad [3.29]$$

The two flows Φ_1 and Φ_2 can be written as:

$$\Phi_1 = \frac{T_1 - T_2}{R_{G1}} = \frac{T_1 - T_2}{R_{GI1} + R_{GC} + R_{GI1}} \qquad [3.30]$$

$$\Phi_2 = \frac{T_1 - T_2}{R_{G2}} = \frac{T_1 - T_2}{R_{GI1} + R_{GB} + R_{GI1}} \qquad [3.31]$$

The overall flow Φ traversing the complete wall is then given by:

$$\Phi = \Phi_1 + \Phi_2 = \frac{T_1 - T_2}{R_{G1}} + \frac{T_1 - T_2}{R_{G2}} = \frac{T_1 - T_2}{R_G} \qquad [3.32]$$

A simple calculation shows that:

$$R_G = \frac{R_{G1} + R_{G2}}{R_{G1}\, R_{G2}} \qquad [3.33]$$

3.1.1.4. *Electrical analogy*

We note that R_{G1} and R_{G2} are related to the overall resistances that make them up, like resistances placed in series. Furthermore, the two expressions for Φ_1 and Φ_2 are related by Ohm's law analogy to temperatures.

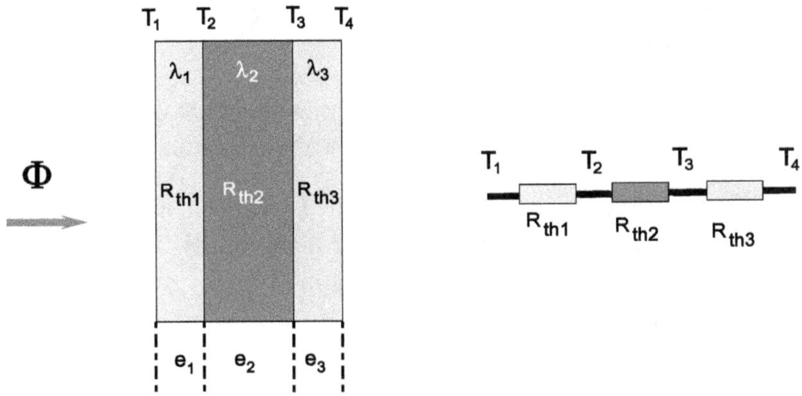

Figure 3.1. *Thermal resistance. Electrical analogy for three superimposed walls. For a color version of this figure, see www.iste.co.uk/ledoux/heat1.zip*

We can then imagine an equivalent electrical diagram, which may induce more complex reasoning when the number of layers in the wall increases.

REMARK.– This method is an approximation. We effectively neglect the fact that the temperature profiles are different in the two superimposed wall systems 1 and 2. The approximation becomes rough if the thermal conductivities of the materials in these two walls become very different.

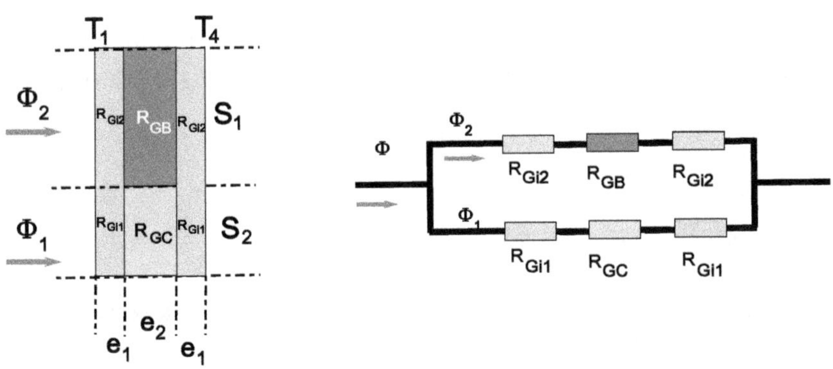

Figure 3.2. *Thermal resistance. Electrical analogy for a composite wall. For a color version of this figure, see www.iste.co.uk/ledoux/heat1.zip*

3.1.2. *Thermal resistance: axisymmetric geometry. The case of a cylindrical wall*

This problem corresponds to the common practical case of a cylindrical duct surrounding a tube that needs to be thermally insulated.

We have a ring-shaped duct of length L with an internal and external radius, made of a material with thermal conductibility λ. We denote T_i as the temperature of the internal face with radius r_i, and T_e as the temperature of the external face with radius r_e.

The problem is a little more complex than the previous one. The single-dimensional heat equation must be solved: we work in cylindrical coordinates (r, θ, z) and the temperatures vary only with the radius r. The equation was given in Chapter 2, section 2.2.2:

$$\frac{d}{dr}\left(\lambda r \frac{dT}{dr}\right) = 0 \qquad [3.34]$$

and for a homogeneous material (which does not vary):

$$\frac{d}{dr}\left(r \frac{dT}{dr}\right) = 0 \qquad [3.35]$$

This second-order equation must be solved with two limit conditions. The temperatures on the two internal and external faces of the ring shape are given by:

$$r = r_i \quad ; \quad T = T_i \qquad [3.36a]$$

$$r = r_e \quad ; \quad T = T_e \qquad [3.36b]$$

We will integrate the differential equation twice in succession, where two constants C_1 and C_2 will emerge:

first integration:

$$r\frac{dT}{dr} = C_1 \qquad [3.37]$$

So $$\frac{dT}{dr}=\frac{C_1}{r} \quad [3.38]$$

second integration:

$$T(r)=C_1\, Ln\, r + C_2 \quad [3.39]$$

The two conditions at the limits will give the value of the constants:

$$T_i = C_1\, Ln\, r_i + C_2 \quad [3.40]$$

$$T_e = C_1\, Ln\, r_e + C_2 \quad [3.41]$$

$$C_1 = \frac{T_i - T_e}{Ln\frac{r_i}{r_e}} = -\frac{T_i - T_e}{Ln\frac{r_e}{r_i}} \quad [3.42]$$

The value of C_2 is not very important because to obtain the flow density, we derive the flow per meter of length and the thermal resistance:

$$\varphi = -\lambda\frac{dT}{dr} = \frac{C_1}{r} = -\lambda\frac{T_i - T_e}{Ln\frac{r_e}{r_i}}\frac{1}{r} \quad [3.43]$$

As we note, the flow density is not the same on both faces (the flow is the same and the lateral surface is different):

internal face:

$$\varphi_i = -\lambda\frac{T_i - T_e}{Ln\frac{r_e}{r_i}}\frac{1}{r_i} \quad [3.44]$$

external face:

$$\varphi_e = -\lambda\frac{T_i - T_e}{Ln\frac{r_e}{r_i}}\frac{1}{r_e} \quad [3.45]$$

The flow on a unit sleeve length will give (fortunately) the same result on both faces:

$$\Phi_i = 2\pi r_i \varphi_i = 2\pi r_i \lambda \frac{T_i - T_e}{Ln\frac{r_e}{r_i}} \frac{1}{r_i} = 2\pi\lambda \frac{T_i - T_e}{Ln\frac{r_e}{r_i}} \qquad [3.46]$$

$$\Phi_e = 2\pi r_e \varphi_i = 2\pi r_e \lambda \frac{T_i - T_e}{Ln\frac{r_e}{r_i}} \frac{1}{r_e} = 2\pi\lambda \frac{T_i - T_e}{Ln\frac{r_e}{r_i}} \qquad [3.47]$$

We therefore introduce here ***a thermal resistance "per unit length"***, a concept that obviously must not be confused with the definitions in plane problems:

$$\Phi = \Phi_i = \Phi_e = 2\pi\lambda \frac{T_i - T_e}{Ln\frac{r_e}{r_i}} = \frac{T_i - T_e}{\frac{Ln\frac{r_e}{r_i}}{2\pi\lambda}} = \frac{T_i - T_e}{R_{th/m}} \qquad [3.48]$$

$$R_{th/m} = \frac{Ln\frac{r_e}{r_i}}{2\pi\lambda} \qquad [3.49]$$

This expression must be remembered while remaining fully aware of the definition of this resistance. We note that its dimension is the inverse of thermal conductibility, which is expressed in $m\,K\,W^{-1}$.

For a homogeneous sleeve of insulation of length L, in the presence of pure conduction, we end up with a flow given by:

$$\Phi = L\frac{T_i - T_e}{R_{th/m}} \qquad [3.50]$$

3.1.3. *Thermal resistance to convection*

In practice, we note a temperature imposed on the face of a wall less often than forced or natural convection. For a plane problem, we can express the density of thermal flow using an expression of the form:

$$\varphi_w = h\,(T_w - T_e) \tag{3.51}$$

where index w denotes the face and index e an "atmospheric" temperature, which is often known as a "distant" temperature in theories of forced convection.

For a cylindrical problem, the radius r of the exchange surface must be taken into account. We therefore work in flow $\Phi_{w/m}$ per unit length (and not in flow density, per m^2):

$$\Phi_{w/m} = 2\pi\, r\, h\,(T_w - T_e) \tag{3.52}$$

3.1.3.1. *Resistance to convection in plane geometry*

We can rewrite φ_w in the form:

$$\varphi_w = h\,(T_w - T_e) = \frac{T_w - T_e}{\dfrac{1}{h}} = \frac{T_w - T_e}{R_{thC}} \tag{3.53}$$

We then note a thermal resistance to convection that is given by (per m^2):

$$R_{thC} = \frac{1}{h} \tag{3.54}$$

Obviously, this resistance will be additive to the resistances of each layer of a plane wall.

3.1.3.2. *Resistance to convection in axisymmetric geometry*

We can rewrite φ_w in the form:

$$\Phi_{w/m} = 2\pi\, r\, h\,(T_w - T_e) = \frac{T_w - T_e}{\dfrac{1}{2\pi\, r\, h}} = \frac{T_w - T_e}{R_{thCL}} \tag{3.55}$$

we have thus defined a thermal resistance per meter of ring:

$$R_{thCL} = \frac{1}{2\pi r\ h} \qquad [3.56]$$

We note that the flow Φ_{wL} that escapes by convection (or that enters or exits from the inside) from a length L of tube will thus be expressed by

$$\Phi_{wL} = L\ \frac{T_w - T_e}{R_{thCL}} \qquad [3.57]$$

A calculation with an analogous form to the plane case will lead to an additivity of these resistances to transfer, both conductive and convective:

$$R_{th/m} = \frac{1}{2\,\pi\, r_i\ h_i} + \sum_{rings} \frac{Ln\left(\dfrac{r_{ext}}{r_{\text{int}}}\right)}{2\,\pi\,\lambda} + \frac{1}{2\,\pi\, r_e\ h_e} \qquad [3.58]$$

In this expression, we take into account the fact that the internal and external convection coefficients can differ (possibly significantly if we have a transfer to a liquid on the inside and to the air on the outside).

REMARK.– in the above formula, it goes without saying that, in the sum of all the *rings*, we attribute the value of internal and external radii of each element to r_e and r_i (see examples for a better understanding).

Thus, a flow that passes through a multilayer sleeve of length L, with internal and external convection, will be expressed by:

$$\Phi = L\frac{T_i - T_e}{R_{th/m}} \qquad [3.59]$$

This will obviously be the same on the internal and external faces.

3.1.4. *Critical radius*

Let us consider a cylinder of radius r_i surrounded by an insulating sleeve. We denote λ as the conductibility of the sleeve insulation and r its external radius. The part of the sleeve with this radius is subject to the convection of coefficient h.

In other words, R_{th} is the thermal resistance of the sleeve, resulting from the resistance of the insulation and the resistance of convection.

We would naturally think that increasing radius r would lead to an increase in insulation. However, the lateral surface of the assembly is also a possible factor of an increase in losses. To study the competition between these two phenomena, we will study the sign of the derivative with respect to r for the resistance R_{th}.

We know that R_{th} has the expression for the case described here:

$$R_{th/m} = \frac{Ln\left(\frac{r}{r_i}\right)}{2\,\pi\,\lambda} + \frac{1}{2\,\pi\,r\,h} \qquad [3.60]$$

and

$$\frac{\partial R_{th/m}}{\partial r} = \frac{\partial}{\partial r}\left[\frac{Ln\left(\frac{r}{r_i}\right)}{2\,\pi\,\lambda} + \frac{1}{2\,\pi\,r\,h}\right] = \frac{1}{2\,\pi\,\lambda\,r} - \frac{1}{2\,\pi\,r^2\,h}$$
$$= \frac{1}{2\,\pi r}\left[\frac{1}{\lambda} - \frac{1}{r\,h}\right] \qquad [3.61]$$

In terms of signs:

$$sign\left[\frac{Ln\left(\frac{r}{r_i}\right)}{2\,\pi\,\lambda}+\frac{1}{2\,\pi\,r\,h}\right]=sign\frac{1}{2\,\pi r}\left[\frac{1}{\lambda}-\frac{1}{r\,h}\right]$$

$$=sign\frac{r\,h-\lambda}{\lambda rh}=sign\frac{r-\dfrac{\lambda}{h}}{\lambda r} \qquad [3.62]$$

by introducing the "critical" radius $r_C=\frac{\lambda}{h}$, we note that:

$$sign\left[\frac{Ln\left(\frac{r}{r_i}\right)}{2\,\pi\,\lambda}+\frac{1}{2\,\pi\,r\,h}\right]=sign\frac{r-r_C}{\lambda r}=sign\left(r-r_C\right) \qquad [3.63]$$

In conclusion, we note that:

– when $r<r_C$, the derivative $\frac{\partial R_{th}}{\partial r}$ is negative. **Increasing** r **leads to a reduction in** R_{th}. **Increasing the thickness of insulation leads to an increase in losses**. This is only paradoxical if we forget that it is in competition with the increase in the lateral surface area;

– when $r>r_C$, the derivative $\frac{\partial R_{th}}{\partial r}$ is **positive. Increasing** r **leads to an increase in** R_{th}. **Increasing the thickness of insulation leads to a reduction in losses**. The insulation takes precedence over the increase in lateral surface area.

This phenomenon is important in practice when we set up the external diameter of an insulating sleeve.

3.2. Examples of the application of thermal resistance in plane geometry

EXAMPLE 3.3.– Heating an igloo

A polar explorer is invited by the Inuit into their igloo. The explorer, a trained physicist then ponders the thermal behavior of this type of building.

REMARK.– The Inuit is the name given to those commonly referred to as Eskimos. This translates simply to 'people'.

The igloo is built of ice. The thickness of the walls is $e = 40\,cm$. The total surface area of the walls of the igloo is $S = 25\,m^2$.

The external temperature is $T_e = -40°\,C$.

We will take into account a convection coefficient h for both inside and outside the igloo.

Figure 3.3. *An igloo (source: chalettime.com). For a color version of this figure, see www.iste.co.uk/ledoux/heat1.zip*

We want to maintain an internal temperature equal to $T_i = 8\,°C$.

REMARK.– The Inuit coat themselves in seal fat for better thermal protection. They also sleep fully clothed. Polar exploration means having to change Western habits.

What thermal power P must the wood fire inside the igloo produce?

We will make the following approximations:

– all calculations assume that the walls have a practically plane geometry;

– losses through the ground (ice) and the smoke hole will be neglected (the Inuit do not like to suffocate);

– the thermal conductibility of the ice is: $\lambda_G = 2.1\ W.m^{-1}.K^{-1}$;

– the convection coefficient (both internal and external) is: $h = 10 W.m^{-2}.K^{-1}$.

SOLUTION.–

We will evaluate the thermal resistance per m² of wall. This includes a component due to the material and a component due to convection.

$$R_{th} = \frac{e}{\lambda} + \frac{1}{h} + \frac{1}{h} = \frac{0,4}{2.1} + \frac{2}{10} = 0.39\ K\,m^2\,W^{-1} \quad [3.64]$$

The energetic power of the fire, that is, the energy that it liberates per unit of time, must compensate for the thermal losses, meaning the thermal flow resulting from the thermal resistance that has previously been calculated.

Thus:

$$P = \Phi = S\frac{T_i - T_e}{R_{th}} = 25\frac{8+40}{0.39} \quad [3.65]$$

$$P = 3077\,W \quad [3.66]$$

EXAMPLE 3.4.– Thermal resistance of a wall

1) Calculate the thermal resistance per m² of a 70 cm-thick stone wall.

2) Calculate the thermal resistance per m² of a 70 cm-thick stone wall, insulated by a 25 cm-thick layer of rock wool.

3) We impose a temperature difference of 11°C between the two faces; calculate the thermal flow traversing through 85 m^2 of uninsulated wall and insulated wall.

We give:

for the stone:

$$e = 70.10^{-2}\ m$$

$$\lambda = 3\ W.m^{-2}.K^{-1}$$

for the rock wool:

$$e = 25.10^{-2}\ m$$

$$\lambda_{LV} = 0.045\ W.m^{-2}.K^{-1}$$

SOLUTION.–

1) We must first calculate the overall thermal resistance R_{th} of the wall.

The wall is simply made up of a layer of stone. The thermal resistance of each layer is calculated by the equation $R_{th} = \frac{e}{\lambda}$, which is applicable to a homogeneous plane wall.

The resistance to transfer of each convection coefficient will be: $R_{th\,conv} = \frac{1}{h}$

For the stone:

$$e = 70.10^{-2}\ m$$

$$\lambda = 3\ W.m^{-2}.K^{-1}$$

$$R_{wall} = \frac{e}{\lambda} = 0.233\ K.m^2.W^{-1} \quad [3.67]$$

2) For the rock wool:

$$e = 25.10^{-2}\ m$$

$$\lambda_{LV} = 0.045\ W.m^{-2}.K^{-1}$$

$$R_{LV} = \frac{e}{\lambda} = 5.55\ K.m^2.W^{-1} \qquad [3.68]$$

Using the additive nature of the resistances to the transfer, the resulting thermal resistance is therefore:

$$R_{th} = 0.233 + 5.55 = 5.78\ K.m^2.W^{-1} \qquad [3.69]$$

3) The flow of outgoing heat for a surface area S of the wall can be calculated by:

$$\Phi = S\ \phi = S\frac{T_i - T_e}{R_{th}} \qquad [3.70]$$

$$\Phi = 85\frac{11}{5.78} = 161.8W \qquad [3.71]$$

EXAMPLE 3.5.– Great Wall of China

The Great Wall of China has, on average, a height of 7 m and a width of 5 m.

Figure 3.4. *The Great Wall of China. For a color version of this figure, see www.iste.co.uk/ledoux/heat1.zip*

We will consider it to be constructed entirely from stone, with thermal conductivity $\lambda = 3 W.m^{-1}.K^{-1}$.

1) What is the thermal resistance per m² if we do not take the wind into account?

2) Rework the calculation, considering the wind on each side to be represented by a convection coefficient $h = 16 W.m^{-2}.K^{-1}$.

3) What is the interest of a calculation of this type, in your opinion?

SOLUTION.–

1) By definition of the thermal resistance of a plane wall:

$$R_{th} = \frac{e}{\lambda} = \frac{5}{3} = 1.67 W^{-1}\ m^2\,°K \qquad [3.72]$$

2) By definition of the thermal resistance:

$$R_{th} = \frac{\lambda}{e} + \frac{2}{h} = 1.67 + \frac{2}{16} = 1.79\ W^{-1}\ m^2\,°K \qquad [3.73]$$

3) If there is one, for learning purposes, it is to show that a stone wall is a very bad insulator.

EXAMPLE 3.6.– Study of losses by conduction through double glazing

Double glazing is made up of two glass panes separated by a layer of immobile dry air.

The thickness of each pane of the glass is 3.5 mm and the layer of air is 12 mm thick. The thermal conductivity of the glass is $0.7 W.m^{-1}.K^{-1}$ and for the air, it is $0.024 W.m^{-1}.K^{-1}$.

1) For a drop in temperature $\Delta T = 5\,°C$ between the two outside faces of the double glazing, calculate the thermal losses for a pane of dimension 1 m^2.

For this evaluation, we neglect the effect of the convection coefficient on either side of each pane of glass.

2) Compare these thermal losses to those obtained with a single pane of thickness 3.5 mm.

SOLUTION.–

Stationary problem: Thermal balances

1) We will calculate the thermal resistances of the panes

The thermal resistance of a plane wall is defined by

$$R_{th} = \frac{e}{\lambda} \quad [3.74]$$

For window B, we associate two types of face, two panes, of thermal resistance R_{thvV} and a layer of air with thermal resistance R_{thair}.

That is, for the glass:

$$e = 3.5.10^{-3}\ m$$
$$\lambda = 0.7\ W.m^{-2}.K^{-1}$$

$$R_V = \frac{e}{\lambda} = 5.10^{-3}\ K.m^2.W^{-1} \quad [3.75]$$

and for the air:

$$e = 1.2.10^{-2}\ m$$
$$\lambda = 0.024\ W.m^{-2}.K^{-1}$$

$$R_{air} = \frac{e}{\lambda} = 0.5\ K.m^2.W^{-1} \quad [3.76]$$

Applying the additive nature of the resistances to transfer, the thermal resistance of the window is given by:

$$R_{th} = \left(2 * 5.10^{-3}\right) + 0,5 = 0.51\, K.m^2.W^{-1} \qquad [3.77]$$

The thermal losses per m², that is, the density of flow of losses, is given by:

$$\varphi = \frac{\Delta T}{R_{th}} \qquad [3.78]$$

$$\Delta T = 5C \qquad [3.79]$$

$$\varphi = \frac{5}{0.51} = 9.8\, W.m^{-2} \qquad [3.80]$$

2) A single-glazed pane would lead to a loss per square meter of:

$$\varphi = \frac{\Delta T}{R_{thV}} = \frac{5}{5.10^{-3}} = 1000\, W.m^{-2} \qquad [3.81]$$

The advantage of double glazing is obvious here. However, we note that the thickness of the glass chosen for this example is relatively weak from a practical point of view.

EXAMPLE 3.7.– Study of the thermal quality of a window

To construct a French window, we consider three glazing options:

A) A 6 mm-thick single-glazed window;

B) A double-glazed window made up of a layer of air of 1.5 cm, bordered by two 6 mm-thick glass panes;

C) A triple-glazed window, made up of three 6 mm-thick glass panes separating two layers of air, each with a thickness of 1.5 cm.

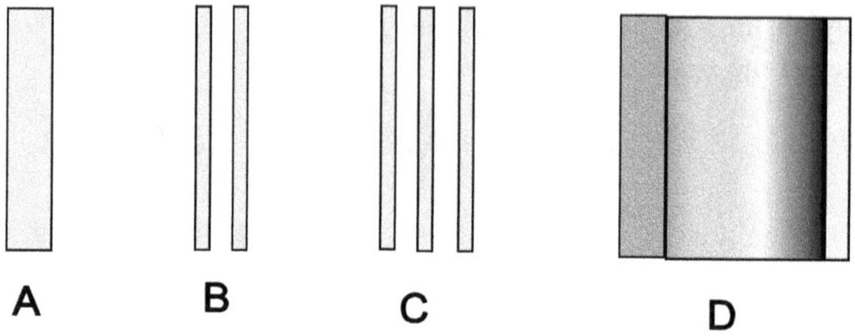

Figure 3.5. *Different types of glazing*

1) Calculate the thermal resistances for 1 m^2 of each of these types of window.

2) Compare these thermal resistances with those for a 20 cm-thick concrete wall (**D**) insulated by a 15 cm-thick layer of rock wool and limited by a layer of BA13 (135 mm-thick plasterboard).

3) Framing of the French window

The triple glazing C is framed in a French window with dimensions given in the following diagram. The frame is made of 4 cm-thick, medium-weight deciduous wood. The dimensions in the diagram are given in cm.

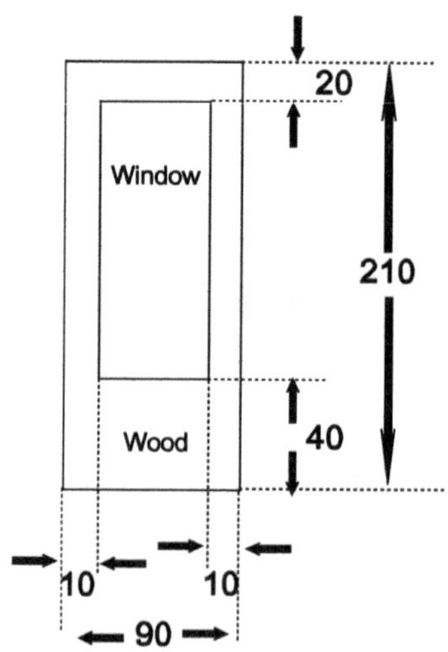

Figure 3.6. *Dimensional drawing of the French window*

The French window is installed in a house. The external temperature is $T_e = 5\,°C$ and the internal temperature is $T_i = 20\,°C$.

Calculate the thermal flow Φ that passes all the way through the French window.

This produces the following thermal conductibilities:

$\lambda_{air} = 0.026\,W.m^{-1}.K^{-1}$ $\lambda_{concrete} = 2\,W.m^{-1}.K^{-1}$

$\lambda_{glass} = 1.05\,W.m^{-1}.K^{-1}$ $\lambda_{BA13} = 0.25\,W.m^{-1}.K^{-1}$

$\lambda_{LV} = 0.041\,W.m^{-1}.K^{-1}$ $\lambda_{wood} = 0.23\,W.m^{-1}.K^{-1}$

SOLUTION.–

1) Calculation of thermal resistances

The thermal resistance of a plane wall is defined by:

$$R_{th} = \frac{e}{\lambda} \qquad [3.82]$$

For window A, we have a single wall with

$$e = 6.10^{-3}\ m$$
$$\lambda = 1.05\,W.m^{-2}.K^{-1}$$

$$R_{thA} = \frac{e}{\lambda} = 5.71.10^{-3}\ K.m^2.W^{-1} \qquad [3.83]$$

For window B, we associate two types of wall.

Two glass panes A, with the thermal resistance $R_{thA} = \frac{e}{\lambda}$, and a layer of air with thermal resistance R_{thair}:

$e = 1.5.10^{-2}\ m$

$\lambda = 0.026 W.m^{-2}.K^{-1}$

$$R_{air} = \frac{e}{\lambda} = 0.577\ K.m^2.W^{-1} \quad [3.84]$$

Applying the additive nature of the thermal resistances “in series”:

$$R_{thB} = 2 * R_{thA} + R_{thair} = 0.588 K.m^2.W^{-1} \quad [3.85]$$

For window C, we associate two types of wall.

Three glass panes A, with thermal resistance R_{thA} , and two layers of air with thermal resistance R_{thair}:

$$R_{thA} = 5.71.10^{-3}\ K.m^2.W^{-1} \quad [3.86]$$

$$R_{thair} = \frac{e}{\lambda} = 0.577\ K.m^2.W^{-1} \quad [3.87]$$

By applying the additive nature of the thermal resistances “in series”:

$$R_{thC} = 3 * R_{thA} + 2 * R_{thair} = 1.171 K.m^2.W^{-1} \quad [3.88]$$

2) Concrete wall

For the concrete wall, we associate three types of wall.

A concrete wall, with thermal resistance $R_{thconcrete} = \frac{e}{\lambda}$, a wall of rock wool with thermal resistance R_{thLV}, and a wall of BA13 plasterboard with the thermal resistance:

$$R_{thBA13} = \frac{e}{\lambda} \quad [3.89]$$

For the concrete:

$$e = 20.10^{-2}\ m$$
$$\lambda = 2\,W.m^{-2}.K^{-1}$$

$$R_{thconcrete} = \frac{e}{\lambda} = 0.1\,K.m^2.W^{-1} \quad [3.90]$$

For the rock wool:

$$e = 15.10^{-2}\ m$$
$$\lambda = 0.041\,W.m^{-2}.K^{-1}$$

$$R_{thLV} = \frac{e}{\lambda} = 3.66\,K.m^2.W^{-1} \quad [3.91]$$

For the BA13:

$$e = 1.3.10^{-2}\ m$$
$$\lambda = 0.25\,W.m^{-2}.K^{-1}$$

$$R_{thBA13} = \frac{e}{\lambda} = 5.2.10^{-2}\ K.m^2.W^{-1}$$

By applying the additive nature of the thermal resistances "in series":

$$R_{thB} = R_{thconcrete} + R_{thLV} + R_{thBA13} = 0.1 + 3.66 + 5.2.10^{-2} = 3.81\,K.m^2.W^{-1} \quad [3.92]$$

3) French window

This flow will be equal to the sum of the flows traversing the wood and the window.

The total surface area of the door is:

$$S = 2.1 * 0.9 = 1.89\,m^2 \quad [3.93]$$

The surface area of the glass S_V through which the flow passes is:

$$S_V = (2.1 - 0.6) * (0.9 - 0.2) = 1.05\, m^2 \quad [3.94]$$

The surface area of wood S_B through which the flow passes is equal to:

$$S_B = 1.89 - 1.05 = 0.84\, m^2 \quad [3.95]$$

The density of flow φ_{Bois} passing through the wood will be:

$$\varphi_{Wood} = \frac{\Delta T}{R_{thwood}} \quad [3.96]$$

with $\Delta T = 20 - 5 = 15\,°C$

and

$$e = 4.10^{-2}\, m$$
$$\lambda = 0.23\, W.m^{-2}.K^{-1}$$

$$R_{thwood} = \frac{e}{\lambda} = 0.174\, K.m^2.W^{-1} \quad [3.97]$$

Thus:

$$\varphi_{wood} = \frac{15}{0.174} = 0.8621\, W.m^{-2} \quad [3.98]$$

The density of flow ϕ_V passing through the glass will be:

$$\varphi_{glass} = \frac{\Delta T}{R_{thglass}} \quad [3.99]$$

with

$$\Delta T = 20 - 5 = 15\,°C \quad [3.100]$$

and (window C):

$$R_{thglass} = 1.171\, K.m^2.W^{-1} \quad [3.101]$$

Thus:

$$\varphi_{glass} = \frac{15}{1.171} = 12.81 W.m^{-2} \quad [3.102]$$

The total flow will be:

$$\Phi = S_B\ \varphi_{wood} + S_V\ \varphi_{glass} \quad [3.103]$$

$$\Phi = (0.84 * 86.2) + (1.05 * 12.81) = 85.86 W \quad [3.104]$$

EXAMPLE 3.8.– Thermal balance of an uninsulated roof

A slate has a thickness of 3 mm.

1) What is its thermal resistance per m²?

2) This slate is mounted on battens, meaning that it is placed on a plank of resinous wood with thickness e = 15 mm.

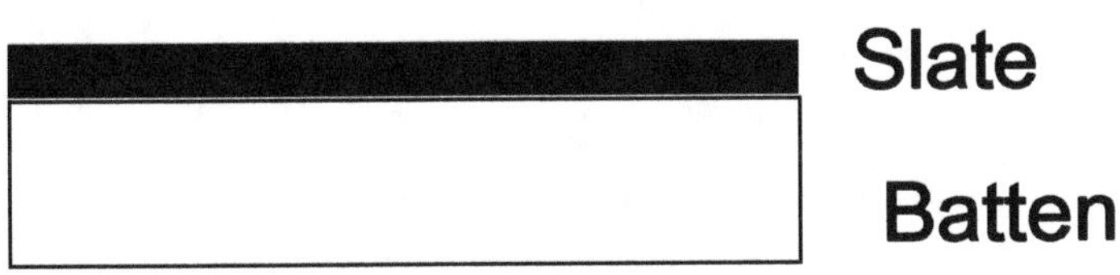

Figure 3.7. *Slate mounted on a batten*

What is the thermal resistance per m² of the slate–batten assembly?

3) A roof is made up of two panes of 20*7 m^2. This roof is made up of battens covered with slates over an attic that is not thermally insulated.

The temperature of the upper face of the slates is $T_e = 30\,°C$, and the temperature of the attic is $T_i = 10\,°C$.

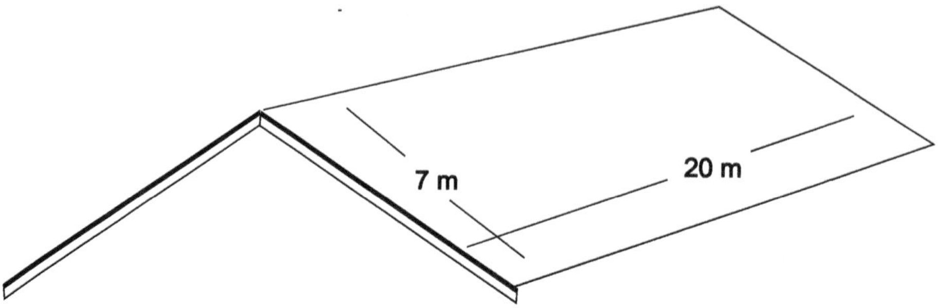

Figure 3.8. *Dimensional drawing of the roof*

What is the flow of heat Φ passing through the roof?

The thermal conductibility of the slate is $\lambda_{slate} = 2.2\,W.m^{-1}.K^{-1}$.

The thermal conductibility of the batten is $\lambda_{batt} = 0.12\,W.m^{-1}.K^{-1}$.

SOLUTION.–

1) The calculation is classic

$$e_{slate} = 3.10^{-3}\,m$$
$$\lambda_{slate} = 2.2\,W.m^{-2}.K^{-1}$$

$$R_{thslate} = \frac{e}{\lambda} = 1.363.10^{-3}\,K.m^2.W^{-1} \quad [3.105]$$

2) Slate on battens

For window B, we associate two types of wall.

The slate, with thermal resistance R_{tha} = $1.363.10^{-3}\,K.m^2.W^{-1}$ and the batten with thermal resistance R_{thvol}:

$$e = 1.5.10^{-2}\,m$$
$$\lambda = 0.12\,W.m^{-2}.K^{-1}$$

$$R_{thbatt} = \frac{e}{\lambda} = 0.125\,K.m^2.W^{-1} \quad [3.106]$$

By applying the additivity of the thermal resistances "in series":

$$R_{thB} = R_{thslate} + R_{thbatt} = 0.1264 K.m^2.W^{-1} \quad [3.107]$$

3) Thermal leaks through the roof

The surface area S through which the heat passes is:

$$S = 2*(20*7) = 280\,m^2 \quad [3.108]$$

The flow density is

$$\varphi_B = \frac{\Delta T}{R_{thB}} \quad [3.109]$$

$$\Delta T = 30 - 10 = 20\,°C$$

$$\varphi_B = \frac{20}{0.1264} = 158.23\,W.m^{-2} \quad [3.110]$$

The total flux is:

$$\Phi = S\varphi_B \quad [3.111]$$

$$\Phi = 280 * 158.23 = 44304\,W \quad [3.112]$$

EXAMPLE 3.9.– Heating a detached house

A detached house has dimensions of 10 m × 7 m on the ground and a height of $H = 3m$.

Its walls are made of agglomerated concrete with a thickness $e = 20\,cm$; they are insulated by 15 cm of rock wool.

The internal h_i and external $H = 3m$ convection coefficients are both equal to $h_i = h_e = 10 W.m^{-2}.K^{-1}$.

The internal temperature is $T_i = 21°C$, and the external temperature is $T_e = 3°C$.

We consider the roof to be very well insulated.

1) What is the total thermal resistance per m²?

2) What is the total flow passing through the walls?

3) What is the power of the radiator heating this detached house?

SOLUTION.–

1) For the concrete:

$$e = 20.10^{-2}\ m$$

$$\lambda = 1.4\ W.m^{-2}.K^{-1}$$

$$R_{concrete} = \frac{e}{\lambda} = 0.143\ K.m^2.W^{-1} \quad [3.113]$$

For the rock wool:

$$e = 15.10^{-2}\ m$$

$$\lambda_{LV} = 0.045\ W.m^{-2}.K^{-1}$$

$$R_{LV} = \frac{e}{\lambda} = 3.33\ K.m^2.W^{-1} \quad [3.114]$$

Using the additive nature of transfer resistances, the resulting thermal resistance is therefore:

$$R_{th} = R_{concrete} + R_{LV} + \frac{1}{h_e} + \frac{1}{h_i} = 0.143 + 3.33 + \frac{2}{10} = 3.67\ K.m^2.W^{-1} \quad [3.115]$$

2) The surface area of the walls is:

$$S = 2 * 17 * 3 = 102\, m^2 \qquad [3.116]$$

The flow of the outgoing heat for a surface area S of the wall can be calculated by:

$$\Phi = S \frac{T_i - T_e}{R_{th}} \qquad [3.117]$$

$$\Phi = 102 \frac{21 - 3}{3.67} = 500.3\ W \qquad [3.118]$$

3) The heating power required to maintain the temperature will be equal to the quantity of heat to introduce into the detached house in units of seconds, in order to compensate for the leaks. It will therefore be equal to the flow of the losses:

$$P = \Phi = 500.3\, W \qquad [3.119]$$

3.3. Examples of the application of the thermal resistance in cylindrical geometry

EXAMPLE 3.10.– Insulating sleeve

1) What is the thermal resistance per meter, R_{th1}, of a rock wool sleeve with an internal diameter $d_i = 6\, cm$ and an external diameter $d_e = 10\, cm$. The thermal conductibility of the rock wool is $\lambda = 0.045\, W.m^{-1}.K^{-1}$.

2) We subject this to a temperature difference of $\Delta T = 25\, °C$ and do not take into account the convection. What is the flow Φ_1 passing through a length $L = 15\, m$ of the sleeve?

3) We take into account the two convection coefficients: an internal coefficient $h_i = 5\, W\, m^{-2}\, K^{-1}$ and an external coefficient $h_e = 8\, W\, m^{-2}\, K^{-1}$. Given these conditions, recalculate the flow Φ_2 through 15 m of this sleeve.

SOLUTION.–

1) The resistance (per meter) of the sleeve is given by the following equation:

$$R_{th1} = \frac{Ln\frac{d_e}{d_i}}{2\pi\lambda} = \frac{Ln\frac{10}{6}}{2\pi * 0.045} = 1.807\, m\,K\,W^{-1} \quad [3.120]$$

2) Similarly, the flow is calculated by:

$$\Phi_1 = L\frac{\Delta T}{R_{th1}} = 15\frac{25}{1.807} = 207.5\,W \quad [3.121]$$

3) We will recalculate the thermal resistance per meter of length:

$$R_{th2} = \frac{Ln\frac{d_e}{d_i}}{2\pi\lambda} + \frac{1}{2\pi\, r_i\, h_i} + \frac{1}{2\pi(r_i + e)h_e}$$

$$= \frac{Ln\frac{10}{6}}{2\pi * 0.045} + \frac{1}{2\pi * 0.01 * 5} + \frac{1}{2\pi * 0.015 * 8} = 6.316\, m\,K\,W^{-1} \quad [3.122]$$

$$\Phi_2 = L\frac{\Delta T}{R_{th2}} = 15\frac{25}{6.316} = 59.37\,W \quad [3.123]$$

EXAMPLE 3.11.– Insulation of a metal tube

A horizontal metal tube with diameter $d_i = 3\,cm$ is maintained at a temperature $T_i = 200°C$. The atmospheric temperature is $T_e = 20°C$.

We want to reduce the flow of heat transferred by this tube to the room in which it is found.

To do this, we surround it with a sleeve of a rock wool insulator of diameter $d_m = 50\,cm$ and thermal conductibility $\lambda = 0.045\,W.m^{-1}.K^{-1}$.

We consider the natural convection coefficient at the external surface of the insulator to be $h = 10\, W\, m^{-2}\, K^{-1}$. We consider that this convection coefficient maintains a very similar value when the sleeve is removed, and that the thermal exchange takes place on the external surface of the tube.

1) Suppose that the transfers are reduced to convection only, what flow of heat is transferred to the environment by a meter of tube if we do not install the insulating sleeve?

2) What is the thermal resistance per meter of length of the sleeve/ convection assembly?

3) What is the flow of thermal loss Φ_2 per meter of length of the tube if the sleeve is installed? Does the insulation solution seem efficient to you?

4) Taking into account the relationship between the flow Φ_2, the atmospheric temperature and the external temperature of the insulating sleeve and of the flow determined in question 3, what is the temperature T_W of the external surface of the sleeve?

SOLUTION.–

1) **We have a simple convection flow**, whose thermal resistance is (per meter):

$$R_{th1} = \frac{1}{2\,\pi\, r_i\, h} = \frac{1}{2\pi * 0.015 * 10} = 1.061\, m\, K\, W^{-1} \qquad [3.124]$$

The flow is calculated as:

$$\Phi_1 = \frac{\Delta T}{R_{th1}} = \frac{200 - 20}{1.061} = 170\, W \qquad [3.125]$$

REMARK.– We could have calculated this flow directly using the definition of the convection coefficient. Calculating the flow through a lateral surface of the tube of 1 m in length:

$$\Phi_1 = S\, h \Delta T = \pi\, d_i\, h \Delta T$$

This obviously gives the same result, since the thermal resistance is defined using this last expression!

2) The outside radius changes and becomes:

$$r_e = \frac{d}{2} = 0.25\, m \qquad [3.126]$$

The thermal resistance takes into account the insulator and the convection:

$$R_{th2} = \frac{Ln\left(\frac{r_e}{r_i}\right)}{2\,\pi\,\lambda} + \frac{1}{2\,\pi\, r_e\, h} = \frac{Ln\left(\frac{0.25_e}{0.015}\right)}{2\,\pi * 0.045} + \frac{1}{2\,\pi\, 0.25 * 10} \qquad [3.127]$$

$$R_{th} = 10.01\, m\, K\, W^{-1} \qquad [3.128]$$

3) The flow becomes:

$$\Phi_2 = \frac{\Delta T}{R_{th2}} = 10.01 = 17.98\, W \qquad [3.129]$$

In other words, there is a reduction in thermal losses of 89%. The efficiency is very satisfactory.

4) The temperature at the surface of the sleeve is obtained from the thermal convection resistance, which is already calculated:

$$R_{thconv} = \frac{1}{2\,\pi\, r_e\, h} = \frac{1}{2\pi * 0.25 * 10} = 6.37.10^{-2}\, m\, K\, W^{-1} \qquad [3.130]$$

On the two sides of the convection limit layer, we then have:

$$\Phi_2 = \frac{T_W - T_e}{R_{thconv}} = \frac{T_W - T_e}{6.37.10^{-2}} = 17.98\, W \qquad [3.131]$$

Therefore:

$$T_W - T_e = 1.14\,°C \qquad [3.132]$$

Moreover:

$$T_{We} = 21.4°C \qquad [3.133]$$

EXAMPLE 3.12.– Insulation of a tube with convection

We insulate a tube with an external diameter $d_e = 6\,cm$ and an internal diameter $d_i = 4\,cm$ through which the fluid flows, and which we consider to be uniform and equal to $T_0 = 170\,°C$.

The external temperature is $T_e = 20\,°C$.

We insulate the tube with a sleeve of rock wool of thickness $e = 5\,cm$.

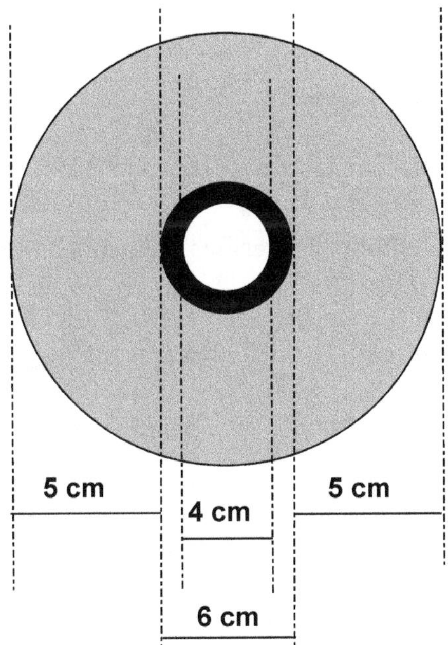

Figure 3.9. *Cross-section through the device*

1) First, we consider the steel tube to be a perfect conductor and the convection is perfect on the outside (with the convection coefficient h being infinite).

a) Calculate the thermal resistance per meter between the fluid and the external atmosphere.

b) Calculate the thermal flow leaving the fluid and transferring to the outside, for a length of 10 m of tube.

2) Second, we take into account an external convection coefficient of $h = 5\,W.m^{-2}.K^{-1}$ and the thermal conductivity λ_{steel} of the tube.

a) Calculate the new value of thermal resistance per meter between the fluid and the external atmosphere.

b) Calculate the new value of the thermal flow leaving the fluid and transferring to the outside over a length of 10 m of tube.

Numerical data

Density of air: $\rho = 1.3\,kg.m^{-3}$.

Specific heat capacity of air: $c_V = J.kg^{-1}$.

The thermal conductibility of various elements and materials is given as follows:

Air: $\lambda_{air} = 0.026\,W.m^{-1}.K^{-1}$ Rock wool: $\lambda_{wool} = 0.047\,W.m^{-1}.K^{-1}$

Glass: $\lambda_{glass} = 1.05\,W.m^{-1}.K^{-1}$ Hard stone: $\lambda_{stone} = 2.4\,W.m^{-1}.K^{-1}$

Steel: $\lambda_{steel} = 54\,W.m^{-1}.K^{-1}$ Wood: $\lambda_{wood} = 0.12\,W.m^{-1}.K^{-1}$

SOLUTION.–

1) The steel tube is a perfect conductor and convection is perfect on the outside.

a) The thermal resistance R_{thLR} per meter of a cylindrical sleeve with an internal radius of r_1 and an external radius of r_2 and with thermal conductibility λ is calculated using the expression:

$$R_{th} = \frac{Ln\left(\frac{r_e}{r_i}\right)}{2\,\pi\,\lambda} \quad [3.134]$$

For the rock wool sleeve, the internal radius is $r_e = \frac{d_e}{2}$ and the external radius is $r_e + e$:

$$R_{thwool} = \frac{Ln\left(\frac{3+5}{3}\right)}{2\,\pi\,0.047} = 3.321\,m^2.K.W^{-1} \quad [3.135]$$

b) The resistance is calculated per meter of tube.

Therefore:

$$\Phi = \frac{\Delta T}{R_{th}} = 10 * \frac{170-20}{3.321} = 451.7\,W \quad [3.136]$$

2) We take into account an external convection coefficient of $h = 5W.m^{-2}.K^{-1}$ and thermal conductivity of λ_{steel} for the tube.

a) The thermal resistance of the tube alone is

$$R_{th} = \frac{Ln\left(\frac{r_e}{r_i}\right)}{2\,\pi\,\lambda} = \frac{Ln\left(\frac{3}{2}\right)}{2*\pi*53} = 1.24.10^{-3}\;m^2.K.W^{-1} \quad [3.137]$$

The thermal resistance (resistance to convective transfer), due to the convection coefficient, is:

$$R_{thconv} = \frac{1}{2\,\pi\,(r_e + e)\,h_e} = 0.398\,m^2.K.W^{-1} \quad [3.138]$$

The resulting thermal resistance is:

$$R_{th} = R_{thube} + R_{thwool} + R_{thconv} \quad [3.139]$$

$$R_{th} = 1.24.10^{-3} + 3.321 + 0.398 = 3.72\, m^2.K.W^{-1} \quad [3.140]$$

b) The thermal flow leaving the fluid and moving to the outside for a 10 m-long tube is calculated in the same way as above with the new thermal resistance:

$$\Phi = \frac{\Delta T}{R_{th}} = 10 * \frac{170 - 20}{3.72} = 403.3\, W \quad [3.141]$$

3.4. Problem of the critical diameter

EXAMPLE 3.13.– Optimization of the insulation of a tube

We have a cylindrical tube with internal radius r_1 and external radius r_2, which is made of a material with thermal conductivity λ_1. Let us assume that we want to insulate with a sleeve with external radius r_3 and thermal conductivity λ_2. Then, h_i and h_e are, respectively, the internal and external convection coefficients.

We will calculate the resistances for a length of 1 m.

1) Give the expression for the thermal resistance R_{thtot} of the tube alone.

2) Give the expression for the thermal resistance R_{thtot} of the assembly tube + sleeve.

3) Determine the conditions for which the addition of a sleeve does indeed lead to a reduction in thermal losses.

Data:

$r_2 = 1.5\, cm$; $\lambda_2 = 0.1\, W.m^{-1}.C^{-1}$; $h_2 = 6\, W.m^{-2}.C^{-1}$

SOLUTION.–

1) Let us calculate the thermal resistance of the tube alone. **We will deal with the problem for 1 m of tube.**

The thermal resistance R_{th} per meter of a cylindrical sleeve, with internal radius r_1, external radius r_2 and thermal conductibility λ, is calculated using the expression:

$$R_{th} = \frac{Ln\left(\frac{r_e}{r_i}\right)}{2\,\pi\,\lambda} \qquad [3.142]$$

NOTE.– We should remember that we can deduce the thermal resistance $R_{th}(L)$ for a sleeve of length L:

$$R_{th}(L) = R_{th}\,L = L\frac{Ln\left(\frac{r_e}{r_i}\right)}{2\,\pi\,\lambda} \qquad [3.143]$$

Applied to the tube alone, we find:

$$R_{th}(L = 1\,m) = \frac{Ln\left(\frac{r_e}{r_i}\right)}{2\,\pi\,\lambda} = \frac{Ln\left(\frac{r_2}{r_1}\right)}{2\,\pi\,\lambda_1} \qquad [3.144]$$

2) The resistance of the sleeve $R_{thsleeve}$ is calculated as above:

$$R_{thsleeve}(L = 1\,m) = \frac{Ln\left(\frac{r_e}{r_i}\right)}{2\,\pi\,\lambda} = \frac{Ln\left(\frac{r_3}{r_2}\right)}{2\,\pi\,\lambda_1} \qquad [3.145]$$

The resistance of a convective layer R_{thconv} is calculated by:

$$R_{thconv} = \frac{1}{2\,\pi\,r\,h} \qquad [3.146]$$

The total resistance results from this:

$$R_{thtot} = \frac{Ln\left(\frac{r_2}{r_1}\right)}{2\pi\lambda_1} + \frac{Ln\left(\frac{r_3}{r_2}\right)}{2\pi\lambda_1} + \frac{1}{2\pi r_1 h_i} + \frac{1}{2\pi r_3 h_e} \qquad [3.147]$$

3) We will examine the effect of adding a sleeve

The parameters r_1 , r_2 , λ_1 and λ_2 are fixed. Therefore, we can only change the values of r_3 and λ_2. We will presume in a first approximation that h_e remains invariable when r_3 varies to a reasonable degree.

The optimization will therefore be related to the component of R_{thtot} that contains r_3, which can be written as:

$$R(r_3) = \frac{Ln\left(\frac{r_3}{r_2}\right)}{2\pi\lambda_1} + \frac{1}{2\pi r_3 h_e} \qquad [3.148]$$

It is therefore necessary to examine the variation of $R(r_3)$ when r_3 varies. We refer to the case studied in section 3.1.4.

We then define the critical radius:

$$r_C = \frac{\lambda_1}{h_e} \qquad [3.149]$$

– when $r_3 < r_C$, the derivative $\frac{\partial R_{th}}{\partial r}$ is negative. An increase in r leads **to a reduction in R_{th}. Increasing the thickness of insulation leads to an increase in losses;**

– when r_3 becomes greater than the critical radius $r_{3c} = \frac{\lambda_1}{h_e}$, it becomes useful to increase the thickness of insulation.

The numerical value of this critical radius is written as:

$$r_2 = 1.5\,cm\,;\ \lambda_2 = 0.1\,W.m^{-1}.C^{-1};\ h_2 = 6\,W.m^{-2}.C^{-1}$$

$$r_{3c} = \frac{0.1}{6} = 1.67\,cm \qquad [3.150]$$

We verify that r_{3c} is indeed greater than r_2. In addition, the increase in the thickness of the sleeve insulation becomes effective when it is greater than $1.67 - 1.5 = 0.17$ cm; in other words, 1.7 mm is low. We see that, in practice, the condition is not very constraining.

EXAMPLE 3.14.– Optimization of the insulation of a steel bar

A steel bar of diameter 5 cm is surrounded by a sleeve of glass with an external diameter of 12 cm. The conductibility of the glass is $\lambda = 1.05\,W.m^{-1}.K^{-1}$. On the outside of the sleeve, the convection coefficient is $h = 8\,W.m^{-2}.K^{-1}$.

The tube is maintained at a temperature $T_B = 70\,°C$. The external temperature is $T_e = 21\,°C$.

1) Calculate the flow of heat between the tube and the external atmosphere for a tube of length 1 m.

2) We replace the glass sleeve of external diameter 12 cm by a new glass sleeve with an external diameter of 26.3 cm. What happens to the heat flow?

3) We replace the glass sleeve of external diameter 26.3 cm by a new glass sleeve with an external diameter of 36 cm. What happens to the heat flow?

4) Compare the flows found in questions **1–3**. What can you observe? How do you explain this result?

SOLUTION.–

1) Sleeve of $D = 12\,cm$

We calculate the thermal resistance R_{th} of the sides of the bottle, using the equation:

$$R_{th} = \frac{Ln\frac{r_e}{r_i}}{2\pi\lambda} = \frac{Ln\frac{r_e}{r_i}}{2\pi\lambda}, \qquad [3.151]$$

which is applicable to a homogeneous cylinder.

$D_i = 5.10^{-2}\,m$

$D_e = 12.10^{-2}\,m$

$\lambda = 1.05\,W.m^{-2}.K^{-1}$

$h_e = 8\,W.m^{-2}.K^{-1}$

$$R_{th} = \frac{Ln\frac{D_e}{D_i}}{2\pi\lambda} = \frac{Ln\frac{12}{5}}{2*\pi*1.05} = 0.133\,W.m^{-2}.K^{-1} \qquad [3.152]$$

The total thermal resistance, including the convection coefficient, is:

$$R_{tot} = \frac{Ln\frac{D_e}{D_i}}{2\pi\lambda} + \frac{1}{2\pi r_e h_e} = 0.133 + \frac{1}{2*\pi*6.10^{-2}*8}$$

$$= 0.133 + 0.3316 = 0.464\,W.m^{-2}.K^{-1} \qquad [3.153]$$

The flow of heat Φ exiting the system per meter for $T_i = T_B = 70°C$ and $T_e = 21°C$ is:

$$\varphi_0 = \frac{T_B - T_e}{R_{tot}} = \frac{70-21}{0.464} = 105.6\,W \qquad [3.154]$$

2) Sleeve of $D = 26.3\,cm$

We calculate the thermal resistance R_{th} of the sides of the bottle, using the equation:

$$R_{th} = \frac{Ln\dfrac{r_e}{r_i}}{2\pi\lambda} = \frac{Ln\dfrac{r_e}{r_i}}{2\pi\lambda} \qquad [3.155]$$

which is applicable to a homogeneous cylinder.

$$D_i = 5.10^{-2}\ m$$
$$D_e = 26.3.10^{-2}\ m$$
$$\lambda = 1.05\ W.m^{-2}.K^{-1}$$
$$h_e = 8W.m^{-2}.K^{-1}$$

$$R_{th} = \frac{Ln\dfrac{D_e}{D_i}}{2\pi\lambda} = \frac{Ln\dfrac{26.3}{5}}{2*\pi*1.05} = 0.251W.m^{-2}.K^{-1} \qquad [3.156]$$

The total thermal resistance, including the convection coefficient, is:

$$R_{tot} = \frac{Ln\dfrac{D_e}{D_i}}{2\pi\lambda} + \frac{1}{2\pi r_e h_e} = 0.133 + \frac{1}{2*\pi*13.1510^{-2}*8}$$

$$= 0.251 + 0.151 = 0.402W.m^{-2}.K^{-1} \qquad [3.157]$$

The flow of heat **Φ** exiting the system per meter for $T_i = T_B = 70°C$ and $T_e = 21°C$ is:

$$\varphi_0 = \frac{T_B - T_e}{R_{tot}} = \frac{70-21}{0.402} = 121.9W \qquad [3.158]$$

3) Sleeve of $D = 36\,cm$

We calculate the thermal resistance R_{th} of the bottle walls, using the equation:

$$R_{th} = \frac{Ln\frac{r_e}{r_i}}{2\pi\lambda} = \frac{Ln\frac{r_e}{r_i}}{2\pi\lambda} \quad [3.159]$$

which is applicable to a homogeneous cylinder.

$D_i = 5.10^{-2}\ m$

$D_e = 36.10^{-2}\ m$

$\lambda = 1.05\ W.m^{-2}.K^{-1}$

$h_e = 8\,W.m^{-2}.K^{-1}$

$$R_{th} = \frac{Ln\frac{D_e}{D_i}}{2\pi\lambda} = \frac{Ln\frac{36}{5}}{2*\pi*1.05} = 0.299\,W.m^{-2}.K^{-1} \quad [3.160]$$

The total thermal resistance, including the convection coefficient, is:

$$R_{tot} = \frac{Ln\frac{D_e}{D_i}}{2\pi\lambda} + \frac{1}{2\pi r_e h_e} = 0.299 + \frac{1}{2*\pi*18.10^{-2}*8}$$

$$= 0.299 + 0.1105 = 0.409\,W.m^{-2}.K^{-1} \quad [3.161]$$

The flow of heat Φ exiting the system per meter for $T_i = T_B = 70°C$ and $T_e = 21°C$ is:

$$\varphi_0 = \frac{T_B - T_e}{R_{tot}} = \frac{70-21}{0.409} = 119.8\,W \quad [3.162]$$

4) Here, we represent the effect of the critical radius on the insulation with a cylindrical sleeve.

The critical radius is

$$r_C = \frac{\lambda}{h_e} = \frac{1.05}{8} = 0.13\,m \qquad [3.163]$$

$$r = 6\,cm < r_C\;;\quad \Phi = 105.6\,W \qquad [3.164]$$

We can then compare with different radii:

$r = 13.15\,cm > r_C$; $\Phi = 121.9W$, the initial radius is smaller than the critical radius, $r = 6\,cm < r_C$, the thickness of the glass increases, and the flow of losses increases.

$r = 18\,cm > r_C$; $\Phi = 119.8\,W$, the initial radius is larger than the critical radius, $r = 13.15\,cm > r_C$, the thickness of glass increases, and the flow of losses decreases.

3.5. Problem with the heat balance

EXAMPLE 3.15. Insulation for a network of tubes

A heat network is made up of steel tubes with an internal diameter $d = 2r_i = 18\,cm$ and an external diameter $d_e = 2r_e = 20\,cm$. The thermal conductibility of the steel is $\lambda_S = 54\;W\,m^{-1}\,K^{-1}$. This tube is lagged with foam of thickness $e = 5\,cm.$ The conductibility of this foam is $\lambda_m = 0.04\;W\,m^{-1}\,K^{-1}$.

Water flowing around the circuit at temperature T_{wat} will be considered, for simplicity, as constant, which is equal to 70°C. The temperature of the ground T_G is homogeneous, which is equal to 6°C. The internal convection coefficient is given by $h_i = 700\,W\,m^{-2}\,K^{-1}$.

1) Initially, we assume that the conductibility of the steel is infinite. We also presume that the convection does not involve thermal resistance (infinite convection coefficient).

a) Give the expression and the value of the thermal resistance R_{th1} per meter of foam.

b) What then is the thermal loss for 5 km of pipe?

2) We now take into account the thermal conductibility of the steel. Moreover, we consider the influence of the convection coefficient h between the flow of water and the tube.

a) Give the expression and the value of the thermal resistance per meter R_{th2} for the foam, steel and convection elements.

b) What does the thermal loss then become for 5 km of pipe?

3) The volume of water is $q_V = 10\, m^3\, hr^{-1}$. Does the hypothesis of a constant water temperature for 5 km seem reasonable to you for an **evaluation** of the losses?

SOLUTION.–

1) Using the hypotheses in this question, the thermal resistances of the steel duct and convection are considered to be zero.

a) We have a simple problem of conduction between the water at the given temperature and the air, via a thermal resistance R_{th1} of the foam. This thermal resistance, which is, of course, a resistance per meter of pipe, is calculated using the known relationship:

$$R_{th/m} = \frac{Ln\dfrac{r_e + e}{r_e}}{2\pi\,\lambda_m} = \frac{Ln\dfrac{0.1+0.05}{0,1}}{2\pi\,0.04} \quad [3.165]$$

$$R_{th/m} = 1.613\, K\, m\, W^{-1} \quad [3.166]$$

b) This resistance is expressed per meter. The flow loss will therefore be:

$$\Phi = L\frac{\Delta T}{R_{th1}} \quad [3.167]$$

with $\Delta T = 70 - 6 = 64\,C$, and $L = 5000\,m$

$$\Phi = 5000 \frac{64}{1.613} = 1.98.10^5\ W \qquad [3.168]$$

2) We will therefore need to add the resistances per meter for the foam, the steel (which is calculated using an identical equation) and the convection coefficient.

a) There will be a resulting resistance, taking into careful account the various values of the radii, which are different for each sleeve:

$$R_{th2} = \frac{Ln\dfrac{r_e + e}{r_i}}{2\pi\lambda_m} + \frac{Ln\dfrac{r_e}{r_i}}{2\pi\lambda_S} + \frac{1}{2\pi h_i} \qquad [3.169]$$

$$R_{th2} = 1.613 + 3.1.10^{-4} + 2.27.10^{-4} \qquad [3.170]$$

We obtain an almost identical value:

$$R_{th2} = 1.613\ K\,m\,W^{-1} \qquad [3.171]$$

b) The thermal loss will therefore be identical to that previously calculated.

3) We will write a thermal balance between the input and the output of the tube: the difference between the incoming heat flow and the outgoing heat flow, with the flow, is equal to the **thermal** losses.

This gives us:

$$\Phi = \rho\, c\, q_V\, (T_{in} - T_{out}) \qquad [3.172]$$

where $\rho = 1000\,kg.m^{-3}$ is the density of water and $c = 4180\,J.kg^{-1}$ is its specific heat capacity (values which should be known in principle). T_{in} and T_{out} are, respectively, the input and output temperatures of water in the tube:

$$q_V = \frac{10}{3600} = 2.78\ m^3\ s^{-1} \qquad [3.173]$$

$$1.98.10^5 = 1000 * 4180 * (T_{in} - T_{out}) \quad [3.174]$$

$$T_{in} - T_{out} = 16.97\,C \quad [3.175]$$

The loss of 17°C represents 24% of the input temperature. This is quite a rough approximation.

4

Quasi-stationary Model

We often want to know how the temperature of a body, that is subject to an outside transfer (either heating or cooling), will change.

4.1. We can perform a simplified calculation, adopting the following hypotheses

a) The body with volume V_{OL} is spatially homogeneous: in particular, its temperature $T(t)$, volumetric mass ρ and specific heat capacity c are spatially constant.

b) The temperature varies slowly over time: we then suppose that the transfer to the exterior can be calculated using relationships taken from the stationary theory. In addition, the thermal balance, as established, assumes that the temperature is constant.

We can observe that these hypotheses are not sufficient, in several ways:

The *temperature of the body cannot be rigorously constant*, otherwise, transfer to the exterior would not be possible (conduction or convection on internal faces, for example). We then presume that the required temperature gradients are located on a small zone near the internal limits of the body. We could take this into account with a convection coefficient, which is expressed with a thermal resistance (internal and/or external convection).

The transfer should, strictly speaking, correspond to the heat equation in the transitory regime. Here, we effectively neglect the effects of this transitory nature. The timescales applied here must therefore be large.

4.2. Method: instantaneous thermal balance

We resolve the problem with a thermal balance. We write that the variation of the temperature of the body in a very short time dt is due to the transfer with the exterior.

We will give the basis for the calculation here, by way of an example.

REMARK.– Be careful, we cannot learn a "recipe" here, some aspects of the model can vary from one problem to the next. We need to **understand a method**.

So, for a homogeneous body with volume V_{OL}, density ρ and specific heat capacity c, its mass will be:

$$m = \rho V_{OL} \qquad [4.1]$$

During the time dt, the quantity of heat exchanged by the body and the exterior is given by:

$$dQ = \rho \, c V_{OL} \, dT \qquad [4.2]$$

This quantity has the same sign as dT. dQ is therefore positive for heating, and negative for cooling (here, we observe the usual conventions of thermodynamics).

This quantity of heat given or extracted from the body will be due to the thermal transfer, which can generally be written as:

$$\Phi = \frac{T(t) - T_e}{R_{th}} \qquad [4.3]$$

where T_e is the external outside temperature ("distant" from the body).

Written in this way, Φ is positive for a heat transfer that is directed towards the exterior, therefore a loss of heat. Close attention must be paid to this when the balance is established.

During this time, the heat exchange will be Φdt

The balance will therefore be written as follows, taking into account the above equations:

$$d\,Q = -\,\Phi\,dt \qquad [4.4]$$

The *minus* sign is a result of the remark made above. We note that its presence is a result of the choice made when writing $d\,Q$ (choice of the thermodynamic sign convention).

This balance therefore leads us to a differential equation for $T(t)$

$$dQ = \rho\,c\,V_{OL}\,dT = -\Phi\,dt = -\frac{T(t)-T_e}{R_{th}}\,dt \qquad [4.5]$$

$$\frac{dT}{dt} = -\frac{T(t)-T_e}{\rho\,c\,V_{OL}\,R_{th}} \qquad [4.6]$$

This equation must be associated with an initial condition. Often, we write:

$$t = 0\,;\;\; T = T_0 \qquad [4.7]$$

We note that writing R_{th} characterizes a problem. Writing this can, in effect, take into account all or some of the following parameters:

– existence of an insulating layer;

– convection coefficients, external and possibly internal (e.g. the natural convection in a room);

– possible radiation phenomena (which, as we will see in another chapter, can be expressed in certain specific cases by a linear relationship that is analogous to the use of a convection coefficient).

NOTE.– In this type of **approximate treatment**, we make sure that we are choosing constant components for thermal resistance in convection; this is a usual approximation, which is not necessarily always rigorous.

4.3. Resolution

To solve the differential equation, a new unknown function is often introduced:

$$\theta(t)=T(t)-T_e \qquad [4.8]$$

The equation and the initial condition become:

$$\frac{d\theta}{dt}=-\frac{\theta}{\rho\, c V_{OL} R_{th}} \qquad [4.9]$$

$$t=0 \quad ; \quad \theta=T_0-T_e$$

The equation is transformed into the form:

$$\frac{d\theta}{\theta}=-\frac{dt}{\rho_S c R_{th}}=-\alpha\, dt \qquad [4.10]$$

with

$$\alpha=\frac{1}{\rho_S c V_{OL} R_{th}} \qquad [4.11]$$

as the coefficient, which makes the written equation lighter. A source of error that is frequently seen in “copies” in fact comes from errors in the systematic rewriting of slightly complex expressions.

Integrating the two terms of the equation:

$$Ln\,\theta=\ -\alpha t+Ln C \qquad [4.12]$$

A constant $Ln C$ appears, which will be given by the initial condition. The choice of the constant in the form of a logarithm is a clever trick, which allows a change to the exponential of the previous expression to be simplified. Taking the exponential of each term:

$$\exp(Ln\,\theta) = \exp(-\alpha t)\exp(LnC) \qquad [4.13]$$

$$\theta = C\exp(-\alpha t) \qquad [4.14]$$

and

$$t = 0 \qquad T_0 - T_e = C \qquad [4.15]$$

which results in the final expression for the temperature of the bar:

$$T(t) - T_e = (T_0 - T_e)\exp(-\alpha t) \qquad [4.16]$$

This expression is general.

If $T(t) - T_e > 0$, then cooling takes place, so $T(t)$ decreases [4.17a]

If $T(t) - T_e < 0$, then heating takes place, so $T(t)$ increases [4.17b]

As we will repeat, learning this result by heart must be avoided because it can lead to errors, but we must instead be able to carry out the reasoning each time. The following examples will allow the reader to become used to this.

4.4. Applications for plane systems

EXAMPLE 4.1.– Heating of a room by radiation

REMARK.– the radiation manifests itself here by an imposed heat flow term. This example has a natural place in a chapter that is dedicated to conduction, and not to radiation, where the latter field does not require any specialized calculations.

A room in the form of a parallelepiped with dimensions $3\,m \times 4\,m$ on the ground and a height of $2.5\,m$ is lit by a glass window with dimensions $2\,m \times 2\,m$.

We consider that the room is perfectly insulated.

The glass is lit from the instant $t=0$ by a flow of $\Phi=300\,W m^{-2}$. We consider that this flow received by the glass is used entirely to heat the air in the room.

At time $t=0$, the temperature is $T_i=18\,°C$.

1) Write the differential equation which represents the temperature $T_i(t)$ of the room.

2) Solve this equation and give $T_i(t)$

3) How long will the temperature take to reach $T_i=25°C$? How long will the temperature take to reach $T_i=70°C$?

4) Here, the model is very simplified. Under real conditions, do you think that a temperature of $70°C$ can be reached in the room? Why?

SOLUTION.–

1) A simple thermal balance gives the differential equation for $T_i(t)$

$$\rho\, V_{OL}\, c\, dT=\Phi\, S\, dt \qquad [4.18]$$

the notation and values associated with this are classic $\rho_i=1.3\,kg.m^{-1}$, $c=1000\,J.kg^{-1}.K^{-1}$

$$V_{OL}=3*4*2.5=30\,m^3 \text{ and } S=4\,m^2 \text{ is the glass surface} \qquad [4.19]$$

$$\frac{dT}{dt}=\frac{\Phi\, S}{\rho\, V_{OL}\, c} \qquad [4.20]$$

with the initial condition

$$t=0 \quad ; \quad T=T_i=18°C \qquad [4.21]$$

2) Resolution is immediate.

We obtain a linear solution over time:

$$T = \frac{\Phi S}{\rho V_{OL} c} t + Cst \qquad [4.22]$$

The constant Cst is given by the initial condition

$$T(0) = \frac{\Phi S}{\rho V_{OL} c} 0 + Cst = T_i \qquad [4.23]$$

Therefore:

$$T = T_i + \frac{\Phi S}{\rho V_{OL} c} t \qquad [4.24]$$

Numerically:

$$\frac{\Phi S}{\rho V_{OL} c} = \frac{300 * 4}{1.30 * 30 * 1000} = 3.08.10^{-2} \qquad [4.25]$$

$$T = 18 + 3.08.10^{-2}\ t \qquad [4.26]$$

3) We will reach $T = 25°C$ by increasing the temperature by $8°\mathrm{C}$, in other words, time:

$$t_{25} = \frac{8}{3.08.10^{-2}} = 260 s = 4 mn\ 20 s \qquad [4.27]$$

We will reach $T = 70°C$ by increasing the temperature by $52°\mathrm{C}$, in other words, time:

$$t_{25} = \frac{52}{3.08.10^{-2}} = 1688 s = 28 mn\ 8 s \qquad [4.28]$$

4) The numbers are not realistic, above all the second, because at a temperature of $T = 70°C$, it will not be easy to neglect the thermal losses in the building!

EXAMPLE 4.2.– Thermal evolution of a garden hut

Where necessary, we will use the numerical data that is also given at the end of the text.

A small garden hut is made up of a single room with internal dimensions $5*7\,m^2$ on the ground and a ceiling height of $2.5\,m$.

The walls are made of concrete with a thickness of $20\,cm$. The roof is made of a concrete terrace area of the same thickness.

We insulate all the walls and the ceiling with a $25\,cm$ layer of expanded polystyrene and BA13 plasterboard ($13\,mm$ thick).

That is, h_i and h_e are the internal and external convection coefficients respectively (valid for the walls and ceiling).

1) Calculate the thermal resistance per m² of the insulated wall.

2) The garden hut is heated by an electric radiator.

The external temperature is $T_e = 6°C$. What minimum power P_{min} must this radiator have in order to maintain a constant temperature in the garden hut of $T_i = 20°C$?

We will neglect the thermal losses through the ground. This is a calculation for an order of magnitude; we can presume that the doors and windows have the same thermal resistance as the walls.

3) Radiator breakdown.

Initially, this question will be resolved literally; only the numerical applications requested in question four will be completed.

The radiator breaks down. We presume that the thermal losses can still be evaluated, as they can in the stationary regime, when the internal temperature of the garden hut T_i (t) varies with time. The external temperature T_e remains constant.

We will call R_{th} the total thermal resistance of the garden hut, S the total surface area of the walls and V_{ol} the internal volume of the garden hut.

a) Write the expression for the instantaneous flow of thermal losses from the garden hut.

b) Writing the thermal balance for the garden hut, give the differential equation for the behavior of $T_i(t)$. It will be useful to introduce the function $\theta(t) = T_i(t) - T_e$

4) Numerical application.

$T_e = 6\,°C$, as above

$T_{i0} = 20\,°C$

After the breakdown of the radiator, how long did the temperature take to fall to $T_i = 10\,°C$ inside the garden hut? Give this time in hours and minutes.

5) Another insulation solution.

Consider a DIY enthusiast who thinks it would be a good idea to replace the polystyrene and plasterboard on the walls and ceiling, with a sandwich of five BA13 plasterboards, separated by four air blasts of thickness $2\,cm$.

a) Calculate the new thermal resistance per m².

b) T_e and T_{i0} maintain the previous values of 6°C and 20°C. How long, in hours and minutes, will it take for the temperature of the interior of the garden hut to descend to $T_i = 10\,°C$ with this new insulation?

c) Does this second solution appear to be more satisfactory than the previous one?

d) Would it have been more useful to replace the polystyrene by glass wool?

Additional data

Thermal conductibility of air: $\lambda_{air} = 0.026 W.m^{-1}.K^{-1}$

Thermal conductibility of concrete: $\lambda_{concrete} = 1.4 W.m^{-1}.K^{-1}$

Thermal conductibility of polystyrene: $\lambda_{poly} = 0.035 W.m^{-1}.K^{-1}$

Thermal conductibility of plasterboard: $\lambda_{BA13} = 0.25 W.m^{-1}.K^{-1}$

Density of air: $\rho_{air} = 1.3 kg.m^{-3}$

Specific heat capacity at constant pressure of air: $C_p = 1007 J.kg^{-1}$

Specific heat capacity at constant volume of air: $C_v = 719 J.kg^{-1}$

$h_i = h_e = 5 W.m^{-2}.K^{-1}$

SOLUTION.–

1) Thermal resistance of the wall

For the wall, we associate three different types of walls:

– a concrete wall with thermal resistance $R_{thconcrete}$;

– a wall of expanded polystyrene with thermal resistance R_{thPoly};

– a wall of BA13 plasterboard with resistance R_{thBA13}

We will calculate the thermal resistances.

For the concrete:

$$e = 20.10^{-2}\, m$$
$$\lambda = 1.4 W.m^{-2}.K^{-1}$$

$$R_{thconcrete} = \frac{e}{\lambda} = 0.143\, K.m^2.W^{-1} \qquad [4.29]$$

For the polystyrene:

$$e = 25.10^{-2}\, m$$
$$\lambda = 0.035 W.m^{-2}.K^{-1}$$

$$R_{thpoly} = \frac{e}{\lambda} = 7.143 K.m^2.W^{-1}$$

For the BA13 plasterboard:

$$e = 1.3.10^{-2}\, m$$
$$\lambda = 0.25 W.m^{-2}.K^{-1}$$

$$R_{thBA13} = \frac{e}{\lambda} = 5.2.10^{-2}\, K.m^2.W^{-1} \qquad [4.30]$$

Applying the additive nature of the thermal resistances "in series", also taking into account the transfer resistances that result from the convection coefficients:

$$R_{thB} = R_{thbconcrete} + R_{thLV} + R_{thBA13} + \frac{1}{h_{\text{int}}} + \frac{1}{h_{ext}} = 7.74 K.m^2.W^{-1} \qquad [4.31]$$

2) Calculation of the leaks in static

The surface area which the leaks are distributed over is:

$$S = 2*(7+5)*2.5 + 7*5 = 95\, m^2 \qquad [4.32]$$

The volume of the room is:

$$V_{OL} = 7*5*2.5 = 87.5 m^3 \quad [4.33a]$$

The difference in temperature is:

$$\Delta T = 20 - 6 = 14 C \quad [4.33b]$$

The flow of the leaks will be:

$$\Phi = S\varphi = S\frac{\Delta T}{R_{thB}} = 95*\frac{14}{7,17} = 171.83 W \quad [4.34]$$

3) Radiator breakdown

a) We will give a general expression for the instantaneous flow of thermal losses from the garden hut.

$$\Phi = S\frac{T_i(t) - T_e}{R_{thB}} \quad [4.35]$$

b) Equation giving the evolution of $T_i(t)$

During a time period dt, the flow of heat exiting the garden hut Φdt leads to cooling dT_i such that:

$$\Phi\, dt = -\rho C_V V_{OL}\, d T_i \quad [4.36]$$

The minus sign expresses that the outgoing heat leads to the cooling of the room.

Replacing Φ by its expression and introducing θ, we obtain the differential equation:

$$\rho_{air} C_V V_{OL} \frac{dT_i}{dt} = -S\frac{\theta}{R_{thB}} \quad [4.37]$$

This can be rewritten, noting that:

$$\frac{dT_i}{dt} = \frac{d(T_i - T_e)}{dt} = \frac{d\theta}{dt} \quad [4.38]$$

$$\frac{d\theta}{dt} = -\alpha\theta \qquad [4.39]$$

with

$$\alpha = \frac{S}{\rho_{air} C_V V_{OL} R_{thB}} \qquad [4.40]$$

c) Calculation of T_i (t)

This is a first-order linear differential equation. A single limit condition will be necessary.

In other words:

$$\theta_0 = T_{i0} - T_e \qquad [4.41]$$

At $t = 0$; $\theta = \theta_0$ [4.42]

$$\frac{d\theta}{dt} = -\alpha\theta \qquad [4.43]$$

$$\frac{d\theta}{\theta} = -\alpha dt \qquad [4.44]$$

$$d\, Ln\theta = -\alpha t + LnC \qquad [4.45]$$

C is a constant. The choice of a constant in LnC means we can immediately write the following as:

$$\theta(t) = C \exp -\alpha t \qquad [4.46]$$

The condition for initial time immediately gives $C = \theta_0$

$$\theta(t) = \theta_0 \exp -\alpha t \qquad [4.47]$$

4) Numerical application

$T_e = 6°C$, as previously seen

$T_{i0} = 20°C$

We are trying to find how long it takes after the radiator breakdown for the temperature to drop to 10°C inside the garden hut.

For 10°C, we have

$$\theta = 10 - 6 = 4°C \; ; \; \theta_0 = 20 - 6 = 14°C \qquad [4.48]$$

Calculation of α gives

$$\alpha = 1.5.10^{-4} \qquad [4.49]$$

$$\theta(t) = \theta_0 \exp -\alpha t \text{ gives } t = \frac{-1}{\alpha} Ln \frac{\theta}{\theta_0} \qquad [4.50]$$

$$t = \frac{-1}{1.5.10^{-4}} Ln \frac{4}{14} = 8352\,s \qquad [4.51]$$

$$t = 8352\,s = 10h\;19mn\;12s$$

5) Another insulation solution

a) With polystyrene, the new thermal resistance per m² is calculated using the same procedure as above. The thermal resistance per m² of a wall becomes

$$R_{rhB2} = \frac{0.2}{0.2} + 5 * \frac{1.3.10^{-2}}{0.25} + 4 * \frac{2.10^{-2}}{0.026} + \frac{1}{5} + \frac{1}{5} = 4.737\,m^2.K.W^{-1} \qquad [4.52]$$

b) With this new insulation, time is proportional to $\frac{1}{\alpha}$ and therefore proportional to the thermal resistance R_{th}, with all of the other parameters remaining constant.

The new time taken to fall to $10°C$ will therefore be:

$$t = 9232\frac{4.737}{8.595} = 5088\,s = 1h\ 24mn\ 38\,s \quad [4.53]$$

c) With this second solution, the cooling time is twice as weak as for the initial insulation! Obviously, this solution is less satisfactory than the previous one.

d) It would not have been more advantageous to replace the polystyrene with glass wool, since the thermal conductivity $\lambda = 0.041 W.m^{-1}.K^{-1}$ of the glass wool (0.041) is greater than that for polystyrene $\lambda = 0.035 W.m^{-1}.K^{-1}$

EXAMPLE 4.3.– Temperature of a block of metal

We advise you to process questions 1 to 3 literally. Numerical calculations will only be carried out for question 4.

A steel block has a rectangular base with sides a and b and height H. Its specific heat capacity is denoted c.

This parallelepiped is placed on its base. We will estimate that no thermal transfer is possible through the base.

The cube can only transfer energy to the exterior via its other faces, for which a convection coefficient h applies.

In addition, the density of convection flow will be considered homogeneous on all exposed faces of the block.

The initial temperature of the block is T_0, higher than the atmospheric temperature T_e.

1) Give the expression of the flow density φ_1, then of the flow Φ_1 coming out of the block at the initial moment in time.

2) The block cools down. We will note $T(t)$ as its instantaneous temperature that we will presume to be homogeneous. Write the expression for the instantaneous flow $\Phi(t)$ that exits the block as a function of T(t) and the relevant parameters.

Deduce from this, by writing the instantaneous heat balance in the block, the differential equation that describes the evolution of the temperature of the block.

3) Find the expression for $T(t)$. To do so, it will be useful to define a variable $\theta(t)=T(t)-T_e$

4) Numerical applications

We give

$$a=30\,cm\ ;\ b=20\,cm\ ;\ H=15\,cm$$

Density of steel:

$$\rho_{steel}=7864\,kg.m^{-3}$$

Specific heat capacity of steel, $c=460\,J.kg^{-1}.K^{-1}$

$$h=5\,W.m^{-2}.K^{-1}$$

$$T_e=20°C$$

$$T_0=150\,°C$$

a) Give the value of φ_1 and Φ_1

b) How long will it take for the difference between the temperature of the block and the outside to descend to 10% of its initial value?

SOLUTION.–

1) The expression of the flow density φ_1, then of the flow Φ_1 is calculated using the definition of the convection coefficient:

$$S = 2(a+b)H + ab = 0.15 + 0.06 = 0.21 m^2 \quad [4.54]$$

$$V_{OL} = abH = 9.10^{-3}\, m^3 \quad [4.55]$$

$$\varphi_1 = h(T_0 - T_e) = 5.130 = 650 W.m^{-2} \quad [4.56]$$

$$\Phi_1 = Sh(T_0 - T_e) = 0.21*5*130 = 136.5W \quad [4.57]$$

2) The outgoing heat over time dt leads to cooling dT_i:

$$\Phi(t) = Sh\left[T(t) - T_e\right] \quad [4.58]$$

$$\rho_{steel} V_{ol}\, c\, dT_i = -\Phi\, dt \quad [4.59]$$

$$\rho_{steel} V_{ol}\, c \frac{dT_i}{dt} = -\Phi = -Sh\left[T(t) - T_e\right] \quad [4.60]$$

3) This equation is solved immediately.

We define a variable

$$\theta(t) = T(t) - T_e \quad [4.61]$$

$$\theta = (T_i - T_e) \quad [4.62]$$

$$\frac{d\theta}{dt} = -\alpha\theta \quad [4.63]$$

$$\alpha = \frac{Sh}{\rho_{steel} V_{ol}\, c} = 3.225.10^{-5} \quad [4.64]$$

$$\frac{d\theta}{dt} = -\alpha\theta \quad [4.65]$$

$$\frac{d\theta}{\theta} = -\alpha dt \quad [4.66]$$

Integrating the two terms:

$$Ln\theta = -\alpha t + Ln\, C \quad [4.67]$$

$$\theta = C \exp - \alpha t \quad [4.68]$$

$$t = 0 \ ; \ \theta = \theta_0 \quad [4.69]$$

$$\theta = \theta_0 \exp - \alpha t \quad [4.70]$$

4) Numerical applications

a) Values of φ_1 and Φ_1

$$S = 2(a+b)H + ab = 0.15 + 0.06 = 0.21 m^2 \quad [4.71]$$

$$V_{OL} = abH = 9.10^{-3} m^3 \quad [4.72]$$

$$\varphi_1 = h(T_0 - T_e) = 5.130 = 650 W.m^{-2} \quad [4.73]$$

$$\Phi_1 = S h(T_0 - T_e) = 0.21*5*130 = 136.5 W \quad [4.74]$$

b) We want to determine how long it will take to drop to 10% of the initial value

$$\theta = \theta_0 \exp - \alpha t \quad [4.75]$$

We will calculate the corresponding value of

$$\theta_0 = 150 - 20 = 130° \quad [4.76]$$

$$\alpha = 3.225.10^{-5} \quad [4.77]$$

$$t = \frac{-1}{\alpha} \ln \frac{\theta}{\theta_0} \quad [4.78]$$

$$\frac{\theta}{\theta_0} = 0.9 \quad [4.79]$$

$$t = \frac{-1}{\alpha} \ln \frac{\theta}{\theta_0} \quad [4.80]$$

EXAMPLE 4.4.– Life in a castle is not always ideal

An 18th-century castle is built of hard stone. The walls are 75 cm thick.

The plane is rectangular, the facades (see drawing) are respectively 100 m and 40 m in length. The height of the facades is uniform and equal to 20 m.

Each of the two facades of length 100 m are pierced by 100 m^2 of windows. These windows are made of glass that is 8 mm thick.

The total surface area of the walls (except the windows of course) is covered with wood of thickness 8 cm.

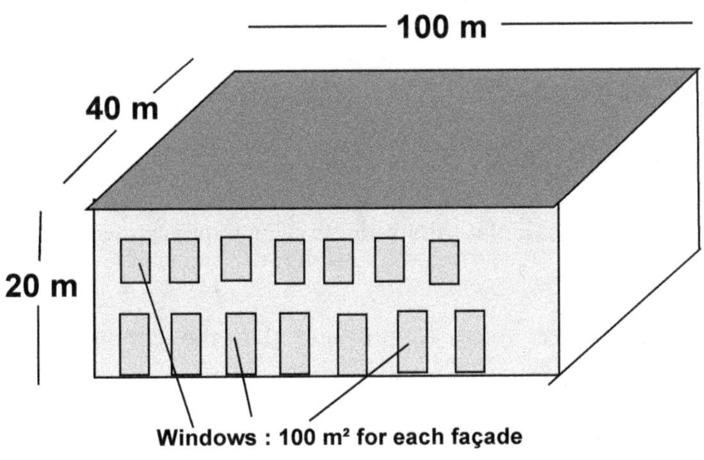

Figure 4.1. *Representation of the castle. For a color version of this figure, see www.iste.co.uk/ledoux/heat1.zip*

We consider two convection coefficients, interior $h_i = 10 W.m^{-2}.K^{-1}$ and exterior $h_e = 10 W.m^{-2}.K^{-1}$; these coefficients apply to the walls, the terrace and the windows.

In this highly simplified model, we will neglect the thermal losses through the roof.

1) Calculate the thermal resistance (per m²) of the walls and windows R_{thWall} and $R_{thWindows}$, respectively, taking into account the wooden surrounding and the convection. Calculate the respective surface areas of the wall, the terrace and the window.

2) We are in winter. The outside temperature is $T_e = 5°C$. We wish to maintain a temperature of $T_{i0} = 14°C$ inside the castle.

What heating power is required to maintain this temperature of 14°C inside the castle?

3) We have managed to achieve a temperature of $T_{i0} = 14°C$ inside the castle, but there is no more wood to put in the chimneys.

The outside temperature stays at $T_e = 5°C$.

We are going to calculate the reduction in temperature $T_i(t)$ inside the castle.

To do so, we will presume that the conductive transfer is quasi-stationary.

The outside temperature stays at $T_e = 5°C$.

We are going to calculate the reduction in temperature $T_i(t)$ inside the castle.

To do so, we are going to presume that the conductive transfer is quasi-stationary.

a) Write the relationship between the instantaneous thermal loss rate and the instantaneous internal and external temperatures $T_i(t)$ and *Te(t)* respectively.

b) Write the thermal balance equation that connects the derivative of the internal temperature with respect to time and this thermal loss.

c) Resolve the equation and give the expression for $\theta(t)$.

We will define $\theta_0 = T_{i0} - T_e$

d) How long will it take for the temperature to become 6°C inside the castle? Give this time in hours and minutes.

It was cold in Versailles during Louis XIV's era. Can you understand him?

Additional data: Some properties for various materials

$\lambda_{air} = 0.026 W.m^{-1}.K^{-1}$ $\lambda_{rockwool} = 0.047 W.m^{-1}.K^{-1}$

$\lambda_{glass} = 1.05 W.m^{-1}.K^{-1}$ $\lambda_{hardstone} = 2.4 W.m^{-1}.K^{-1}$

$\lambda_{steel} = 54 W.m^{-1}.K^{-1}$ $\lambda_{wood} = 0.12 W.m^{-1}.K^{-1}$

Density of air:

$\rho = 1.3 kg.m^{-3}$

Specific heat capacity of air:

$c_v = 718 J.kg^{-1}.K^{-1}$

SOLUTION.–

1) We will use the additive nature of resistances to transfer.

For the wall, the thermal resistance per m² is:

$$R_{thmur} = \frac{e_{wall}}{\lambda_{stone}} + \frac{e_{wood}}{\lambda_{wood}} + \frac{1}{h_i} + \frac{1}{h_e} \quad [4.81]$$

$$R_{thwall} = 0.312 + 0.667 + 0.1 + 0.1 = 1.179°C \quad [4.82]$$

For glass, the thermal resistance per m² is:

$$R_{thV} = \frac{e_{glass}}{\lambda_{glass}} + \frac{1}{h_i} + \frac{1}{h_e} \quad [4.83]$$

$$R_{thV} = \frac{8.10^{-3}}{1.05} + 0.2 = 0.207\, m^2.K.W^{-1} \quad [4.84]$$

The surface areas are:

$$S_{Total} = 5600\, m^2 \quad [4.85a]$$

$$S_{wall} = 5400\, m^2 \quad [4.85b]$$

$$S_{glass} = 200\, m^2 \quad [4.85c]$$

2) Heating power is necessary to maintain the temperature of 14°C inside the castle.

This power is equal to the thermal flow that escapes by conduction:

$$\Phi = S_{wall} \frac{T_i - T_e}{R_{thwall}} + S_{glass} \frac{T_i - T_e}{R_{thglass}} \quad [4.86]$$

$$\Phi = 5400 \frac{9}{1.179} + 200 \frac{9}{0.207} = 4.992.10^4\, W \quad [4.87]$$

In other words, a power of approximately 50 kW.

3) There is no more wood to burn in the chimneys.

a) We will determine the relationship between the flow of instantaneous heat loss and the instantaneous internal and external temperatures $T_i(t)$ and $T_e = Ct$ respectively.

Application of a quasi-stationary process is equivalent to maintaining the relationship between flows and temperature written in the stationary regime:

$$\Phi = S_{wall} \frac{T_i(t) - T_e}{R_{thwall}} + S_{glass} \frac{T_i(t) - T_e}{R_{thglass}} \qquad [4.88]$$

b) Balance equation between the outgoing flow and cooling of the air in the castle:

$$\Phi\, dt = -\,\rho\, C_V\, V_{OL}\, d\, T_i \qquad [4.89]$$

The minus sign expresses that the outgoing heat leads to the cooling of the room.

Replacing Φ by its expression, introducing the variable θ, we find the differential equation:

$$\theta = T_e - T_i(t) \qquad [4.90]$$

$$\rho_{air}\, C_V\, V_{OL} \frac{dT_i}{dt} = -\,S_{wall} \frac{T_i(t) - T_e}{R_{thwall}} + S_{glass} \frac{T_i(t) - T_e}{R_{thglass}} \qquad [4.91]$$

which can be rewritten, noting that:

$$\frac{dT_i}{dt} = \frac{d(T_i - T_e)}{dt} = \frac{d\theta}{dt} \qquad [4.92]$$

$$\frac{d\theta}{dt} = -\alpha\theta \qquad [4.93]$$

with

$$\alpha = \frac{\dfrac{S_{wall}}{R_{thwall}} + \dfrac{S_{glass}}{R_{thglass}}}{\rho_{air}\, C_V\, V_{OL} R_{th}} \qquad [4.94]$$

The volume of the castle is:

$$V_{OL} = 100 * 40 * 20 = 80000 m^3 \qquad [4.95]$$

α gives

$$\alpha = \frac{\dfrac{S_{wall}}{R_{thwall}} + \dfrac{S_{glass}}{R_{thglass}}}{\rho_{air} C_V V_{OL} R_{th}} = \frac{5546}{1.3 * 718 * 8.10^4} = 7.43.10^{-5} \qquad [4.96]$$

This equation, associated with the limit condition:

At the initial instant

$$t = 0 \quad ; \quad \theta = \theta_0 = T_{int0 -} T_{ie} \qquad [4.97]$$

c) The equation is classic and gives the expression for $\theta(t)$.

We define

$$\theta_0 = T_{i0} - T_e \qquad [4.98]$$

$$\frac{d\theta}{dt} = -\alpha\theta \qquad [4.99]$$

$$\frac{d\theta}{\theta} = -\alpha dt \qquad [4.100]$$

$$d\, Ln\theta = -\alpha t + LnC$$

C is a constant. The choice of a constant in LnC means we can immediately write:

$$\theta(t) = C \exp - \alpha t \qquad [4.101]$$

The condition in initial time immediately gives $C = \theta_0$

$$\theta(t) = \theta_0 \exp - \alpha t^2 \qquad [4.102]$$

d) After how long will it be 6°C in the castle?

The time we are seeking is such that:

$$t = t_f; \quad \theta = \theta_f = T_f - T_e = 6 - 5 = 1 \qquad [4.103]$$

$$\theta_f = \theta_0 \exp{-\alpha t_f} \text{ gives } t_f = \frac{-1}{\alpha} Ln \frac{\theta_f}{\theta_0} \qquad [4.104]$$

$$t_f = \frac{-1}{7.43.10^{-5}} Ln \frac{1}{9} = 29572\,s \qquad [4.105]$$

$$t = 8h\ 12mn\ 52s \qquad [4.106]$$

Taking into account the powers to be applied, the phenomenon must have occurred often.

EXAMPLE 4.5.– Thermal considerations for a farmhouse

An old farmhouse in the form of a parallelepiped has a length on the ground of 16 m and a width of 7 m, over a height of 4 m. That is, S, the total surface area of the vertical walls, and V_{OL} the volume that needs to be heated, will be assimilated to a parallelepiped with dimensions $16*7*4\,\text{m}^3$.

We are in winter. The temperature outside is $T_e = 5°C$. We want to maintain a homogeneous internal temperature of $T_{i0} = 19°C$.

1) In the original construction of the farm, the walls of the farm are made of granite of thickness $e = 80\,cm$. The walls are not insulated.

a) Give the expression and the value of the thermal resistance per m² $R_{th1\,1}$ of the farm wall. Take into account an internal convection coefficient $h_i = 9\,W\,m^{-2}K^{-1}$ h_i and an external convection coefficient $h_e = 16\,W\,m^{-2}K^{-1}$.

b) To estimate the losses through the roof, we will consider that they are added to the losses through the walls and are equal to 40% of the losses **through the vertical walls**.

For the calculation, we will assimilate the door and windows (small surface areas) to sections of wall.

What heating power Φ_1 is required to maintain a temperature $T_{i0} = 19°C$ in these conditions in the farm?

2) We insulate the inside of the vertical walls with a layer of $e_2 = 15\ cm$ of rock wool. We insulate the roof so that the losses through the roof still represent 40% of the losses through the vertical walls.

a) What happens to the thermal resistance per m² R_{th2} of the wall of the farm? (Expression and value). We will still take into account an internal convection coefficient $h_i = 9\, W\, m^{-2} K^{-1}$ h_i and an external convection coefficient $h_e = 16\, W\, m^{-2} K^{-1}$

b) What heating power Φ_2 is now necessary to maintain a temperature T_{i0} = 19°C in these conditions in the farm?

3) The heating breaks down when it is $T_{i0} = 19°C$ inside the farm, and the external temperature is still $T_e = 5°C$.

We are obviously in a case where the house is insulated.

Give the expression for the total loss per unit of time Φ as a function of the internal temperature difference $T_i(t)$ and the external temperature T_e.

We will introduce the variable $\theta = T_i(t) - T_e$

a) Write the differential equation for $\theta(t)$ and α, a carefully-selected unique coefficient.

b) Resolve this equation and give the expressions for $\theta(t)$ and $T_i(t)$

c) Numerical application. How long will the temperature take to reach $T = 8°C$?

Numerical data:

Density of air: $\rho = 1.3\, kg.m^{-3}$

Specific heat capacity of air at a constant volume: $c_V = 716\, J.kg^{-1}.K^{-1}$

Thermal conductibility of air: $\lambda_{air} = 0.026\, W\, m^{-1} K^{-1}$

Thermal conductibility of rock wool: $\lambda_W = 0.045\, W\, m^{-1} K^{-1}$

Thermal conductibility of granite: $\lambda_G = 3\, W\, m^{-1} K^{-1}$

SOLUTION.–

1) The walls are not insulated

a) The thermal resistances of insulation and convection are:

$$R_{th1} = \frac{e}{\lambda_G} + \frac{1}{h_i} + \frac{1}{h_e} \qquad [4.107]$$

$$R_{th1} = 0.267 + 0.111 + 6.25.10^{-2} = 0.44\, K.m^2.W^{-1} \qquad [4.108]$$

b) The exchange surface is:

$$S = 184\, m^2 \qquad [4.109]$$

The flow gives:

$$\Phi_1 = S \frac{T_i - T_e}{R_{th1}} * 1.4, \qquad [4.110]$$

where the coefficient $1,4$ takes into account the losses through the roof.

That is:

$$\Phi_1 = 8196\, W \qquad [4.111]$$

This is a high value for heating a house!

2) The walls are insulated

The thermal resistance is increased by the resistance of the insulating material:

$$R_{th1} = \frac{e}{\lambda_G} + \frac{1}{h_i} + \frac{1}{h_e} + \frac{e_2}{\lambda_W} \quad [4.112a]$$

$$R_{th1} = 0.267 + 0.111 + 6.25.10^{-2} + 3.33 = 3.77\ K.m^2.W^{-1} \quad [4.112b]$$

The flow of losses will then be:

$$\Phi_1 = S\frac{T_i - T_e}{R_{th2}} * 1.4 \quad [4.113]$$

$$\Phi_1 = 956.6\ W \quad [4.114]$$

This is much more reasonable!

3) The equation for the temperature results from a classic thermal balance:

$$\rho_{air} C_V V_{OL} \frac{dT_i}{dt} = -S\frac{T_i(t) - T_e}{R_{th2}} 1.4 \quad [4.115]$$

with the initial condition

$$t = 0 \quad ; \quad T_i = T_{i0} = 19°C : \quad [4.116]$$

a) Introducing $\theta(t)$

$$\frac{d\theta_i}{dt} = -S\frac{\theta_e}{\rho_{air} c_V V_{OL} R_{th2}} 1.4 = -\alpha\theta \quad [4.117]$$

with:

$$\alpha = \frac{1.4S}{\rho_{air} c_V V_{OL} R_{th2}} \quad [4.118]$$

the initial condition becomes:

$$t=0 \quad ; \quad \theta_0 = T_{i0} - T_e = 14°C : \qquad [4.119]$$

b) The solution is classic:

Integrating the two terms of the equation:

$$Ln\,\theta = -\alpha t + Ln C \qquad [4.120]$$

$$\theta = C\exp(-\alpha t)$$

at

$$t = 0 \qquad \theta_0 = C = 14 \qquad [4.121]$$

Hence, the final expression for the temperature:

$$T_i(t) - T_e = (T_{i0} - T_e)\exp(-\alpha t) \qquad [4.122]$$

4) We will calculate the time required to drop to $8°C$.

Then,

$$\theta = 3°C \qquad [4.123]$$

We find:

$$\alpha = 1.63.10^{-4} \qquad [4.124]$$

the time we are looking for is:

$$-\alpha t = Ln\frac{T_i(t) - T_e}{T_{i0} - T} \qquad [4.125]$$

$$t_S = \frac{1}{\alpha}Ln\frac{T_{i0} - T}{T_i(t) - T_e} \qquad [4.126]$$

$$t_S = \frac{1}{\alpha} Ln \frac{14}{3} \quad [4.127]$$

$$t_S = 9450\,s = 2\,hr\ 37\ mn \quad [4.128]$$

EXAMPLE 4.6.– A DIY enthusiast is getting ready for a picnic

A DIY enthusiast has built an insulated box to protect their food for their picnic.

This box takes the form of a rectangular parallelepiped. It is constructed from plywood boards of thickness 18 mm. The thermal insulation is completed by polystyrene sheets of thickness 5 cm, which are stuck to the inside of the box. Assembled in this way, the inside of the box is a parallelepiped with a base of $50 * 30$ cm² and a height of 40 cm.

The box has been filled in a house where the atmospheric temperature was $T_{\text{int}\,0} = 16°C$.

This temperature is maintained in the box until it is taken out for the picnic; it is forgotten outside during a family photo session; the atmospheric temperature is $T_e = 25°C$. The external and internal convection coefficients are $h_e = 5W.m^{-2}.K^{-1}$ and $h_i = 5W.m^{-2}.K^{-1}$ respectively. In addition, no thermal flow will pass through the base during the problem.

1) Calculate the thermal resistance per m² of the walls of the box.

2) When the box is taken out, the atmospheric temperature is still $T_{\text{int}\,0} = 16°C$ in the box. What is the thermal flow entering the box at that moment?

3) Since heat enters the box, the internal temperature $T_{\text{int}}(t)$ increases. What is the expression for the total instantaneous flow entering the box?

4) Writing the thermal balance for the air inside the box. Deduce the differential equation from this, which gives the evolution of $T_{\text{int}}(t)$.

Resolve this equation: to do this, it will be practical to introduce the variable $\theta = T_e - T_{\text{int}}(t)$. Give the expression for $\theta(t)$.

5) We consider that certain foods are in danger above 23°C. What is the maximum authorized time for the photography session? Will the DIY enthusiast be pleased by this first test?

6) The DIY enthusiast then uses their coolbox: they introduce a liter of ice into the box.

We will then apply the following:

– the internal temperature of the air in the box becomes $T_{\text{int}} = 7°C$;

– this temperature is maintained as long as all the ice has not melted;

– all of the heat entering the box by conduction is used to melt the ice, whose fusion temperature is L.

a) Since the external temperature is still $T_e = 25°C$, calculate the thermal flow passing through the walls of the box.

b) For how long can the temperature remain at $T_{\text{int}} = 7°C$ in the box?

Is this system more satisfactory than the previous one?

We give:

Density of air: $\rho = 1.3\, kg.m^{-3}$

Density of ice: $\rho_{ice} = 917\, kg.m^{-3}$

Specific heat capacity of air: $c_V = 716\, kg.m^{-3}$

Thermal conductivities of plywood and polystyrene:

$\lambda_{Polystyrene} = 0.035\, W.m^{-1}.K^{-1}$; $\lambda_{plywood} = 0.15\, W.m^{-1}.K^{-1}$

Latent heat of the ice: $L = 333\, kJ$ is necessary to make 1 kg of ice melt.

SOLUTION.–

1) Thermal resistance of the walls of the box

We are opposite a composite wall. The thermal resistance per unit of surface R_{th} will be the sum of the thermal resistances, which can be calculated using $R_{th}=\frac{e}{\lambda}$ for each of the components of this wall. Convection will be present on each side in the form of an equivalent thermal resistance calculated with $R_{th}=\frac{1}{h}$, where h is a convection coefficient.

Using the above data, we therefore obtain:

$$R_{th}=\frac{18.10^{-3}}{0.15}+\frac{5.10^{-2}}{0.035}+\frac{1}{5}+\frac{1}{5}=1.949\,m^2.K.W^{-1} \quad [4.129]$$

2) Thermal flow entering the box at the initial time instant

The area which the leaks are spread out over is:

$$S=2*(0.5+0.3)*0.4+0.5*0.3=0.79\,m^2 \quad [4.130]$$

The volume of the box is:

$$V_{OL}=0.5*0.3*0.4=6.10^{-2}\,m^3 \quad [4.131]$$

The difference in initial temperature is:

$$\Delta T=25-16=9°C \quad [4.132]$$

The initial flow of leakage will be:

$$\Phi=S\varphi=S\frac{\Delta T}{R_{th}}=0.79*\frac{9}{1.949}=3.648W \quad [4.133]$$

3) The total instantaneous flow entering the box is expressed by:

$$\Phi=S\frac{T_e-T_{\text{int}}(t)}{R_{th}} \quad [4.134]$$

4) Differential equation for the evolution of T_{int} (t)

For a time dt, the flow of heat entering the box Φdt leads to heating dT_i such that:

$$\Phi dt = +\rho C_V V_{OL} d T_i \qquad [4.135]$$

The plus sign expresses that the incoming heat leads to the heating of the room.

Replacing Φ by its expression, introducing θ, we find the differential equation:

$$\rho_{air} C_V V_{OL} \frac{d T_{int}}{dt} = -S \frac{\theta}{R_{th}} \qquad [4.136]$$

This can be rewritten, noting that:

$$\frac{d T_{int}}{dt} = -\frac{d(T_e - T_{int})}{dt} = -\frac{d\theta}{dt} \qquad [4.137]$$

$$\frac{d\theta}{dt} = -\alpha\theta \qquad [4.138]$$

with

$$\alpha = \frac{S}{\rho_{air} C_V V_{OL} R_{th}} \qquad [4.139]$$

α gives

$$\alpha = \frac{0,79}{1.3 * 716 * 6.10^{-2} * 1.949} = 7.26.10^{-3} \qquad [4.140]$$

Solving this first-order linear equation requires a limit condition.

At the initial instant in time:

$$t=0 \quad ; \quad \theta=\theta_0=T_{e}-T_{\text{int}\,0} \tag{4.141}$$

$$\frac{d\theta}{dt}=-\alpha\theta \tag{4.142}$$

$$\frac{d\theta}{\theta}=-\alpha\,dt \tag{4.143}$$

$$d\,Ln\theta=-\alpha t+LnC \tag{4.144}$$

where C is a constant. The choice of a constant in LnC means we can immediately write:

$$\theta(t)=C\exp-\alpha t \tag{4.145}$$

The condition at the initial time immediately leads to

$$C=\theta_0 \tag{4.146}$$

$$\theta(t)=\theta_0\exp-\alpha t \tag{4.147}$$

5) Maximum authorized length of the photography session:

For $T_{\text{int}}=23°C$, we have $\theta=2°C$; $\theta_0=25-16=9°C$ [4.148]

Calculation of α gives $\alpha=7.26.10^{-3}$ [4.149]

$$\theta(t)=\theta_0\exp-\alpha t \text{ gives } t=\frac{-1}{\alpha}Ln\frac{\theta}{\theta_0} \tag{4.150}$$

$$t=\frac{-1}{7.26.10^{-3}}Ln\frac{2}{9}=207\,s=3\,mn\,27\,s \tag{4.151}$$

Such a short duration is obviously unacceptable.

6) Use in an icebox

a) The outside temperature is still 25°C.

The thermal flow passing through the box is calculated for a constant temperature:

$$\Phi = S\frac{T_e - T_{\text{int}}}{R_{th}} = 0.79 * \frac{25-7}{1.949} = 7.296W \quad [4.152]$$

b) As long as the internal temperature is constant, heat makes the ice melt.

There is a mass of:

$$m = \rho ice V_{olice} = 917 * 10^{-3} = 0.917 \quad [4.153]$$

to be melted, which will require an energy of $E = mL$ for this fusion. A heating time t_{fusion} is therefore necessary, such that:

$$\Phi \, t_{fusion} = mL \quad [4.154]$$

In other words:

$$\Phi \, t_{fusion} = \frac{mL}{\Phi} = \frac{0.917 * 333.10^3}{7.296} = 41.85.10^3 \, s = 11h37mn30s \quad [4.155]$$

This time is obviously more acceptable, which easily leaves time for transportation and photographs!

EXAMPLE 4.7.– Before fridges, there were iceboxes

In the second half of the 19th century, before refrigerators were invented, iceboxes were an item of furniture that came in two parts. The upper part held a large block of ice that was regularly changed. Food was kept cool in the lower part. Equipment of this kind was still used in the 1950s.

A cube of ice measuring $a_0 = 20\,cm$ is introduced into an icebox. The cube is of course at a temperature of $T_i = 0°C$, and the atmospheric temperature is $T_e = 10°C$. The heat is brought to the cube by a convection coefficient $h = 10W.m^{-2}.K^{-1}$.

The heat for the fusion of the ice is $L_f = 333.55\,kJ.kg^{-1}$

Figure 4.2. *The icebox, the ancestor of the fridge. For a color version of this figure, see www.iste.co.uk/ledoux/heat1.zip*

1) What is the flow of heat Φ that is transmitted to the cube (neglecting the transfer through the shelf which the ice is placed on)?

2) Taking this value of the flow, calculate a time t_f for the fusion of the ice cube.

3) Will this time be the observed fusion time? If not, will the exact result be higher or lower than that seen here? We will not provide an answer by calculation, but instead through reasoning.

Do you think that the order of magnitude found validates the concept of an icebox?

SOLUTION.–

1) The exchange surface is made up of five faces; there is no transfer through the floor:

$$S = 5a^2 = 5*(0,2)^2 = 0.2m^2 \quad [4.156]$$

$$\Phi = Sh(T_e - T_i) = 0.2*10*10 = 20W \quad [4.157]$$

2) If this flow is maintained throughout the fusion, then we will have a fusion time such that

$$\Phi t_f = mL_f \quad [4.158]$$

where $m = \rho a_0^3$ is the mass of the cube and L_f is the fusion heat of the ice.

This gives

$$t_f = \frac{\rho a_0^3 L_f}{\Phi} \quad [4.159]$$

$$t_f = \frac{917*(0.2)^3*333.55.10^3}{20} = 12.24.10^4 = 34h \quad [4.160]$$

3) This time cannot be exact. Effectively, the exchange surface between the air and the ice reduces as the ice melts. The fusion time will therefore be **greater** than the time found.

We can attempt to calculate this time; we will take the unknown $a(t)$.

For a time dt, the flow of heat entering the ice cube Φdt leads to a fusion of the mass dm

$$\Phi dt = -dm L_f \quad [4.161]$$

dm is connected to da by:

$$m = \rho a^3 \quad [4.162]$$

$$dm = 2\rho a^2 da \qquad [4.163]$$

$$\Phi dt = -2\rho a^2 da L_f \qquad [4.164]$$

The differential equation for this is therefore:

$$\frac{da}{dt} = \frac{-\Phi}{2\rho a^2 L_f} \qquad [4.165]$$

with

$$\Phi = 5a^2 h(T_e - T_i) \qquad [4.166]$$

In other words:

$$\frac{da}{dt} = \frac{-5a^2 h(T_e - T_i)}{2\rho a^2 L_f} = \frac{-5\,h(T_e - T_i)}{2\rho\,L_f} \qquad [4.167]$$

And again

$$\frac{da}{dt} = -\alpha \qquad [4.168]$$

with:

$$\alpha = \frac{5\,h(T_e - T_i)}{2\rho\,L_f} = \frac{5*10*10}{2*917*333.55.10^3} = 8.17.10^{-7} \qquad [4.169]$$

With the limit condition

$$t = 0 \quad ; \quad a = a_0 \qquad [4.170]$$

The solution is obvious and gives:

$$a(t) = a_0 - \alpha t \qquad [4.171]$$

The fusion is complete when $a\left(t_f\right)=0$, in other words, the time t_f:

$$t_f = \frac{a_0}{\alpha} = \frac{0,2}{1.63.10^{-6}} = 2.45.10^5\ s \qquad [4.172]$$

$$t_f = 2.45.10^5\ s = 67\,h\,58\,mn \qquad [4.173]$$

The orders of magnitude found here do indeed validate the concept.

Numerical data:

Densities

Density of air: $\rho = 1.3\,kg.m^{-3}$

Density of agglomerated concrete: $\rho = 1{,}900\,kg.m^{-3}$

Density of stone: $\rho = 2{,}600\,kg.m^{-3}$

Density of steel: $\rho = 7{,}833\,kg.m^{-3}$

Density of ice: $\rho = 917\,kg.m^{-3}$

Density of air: $\rho = 1.3\,kg.m^{-3}$

Specific heat capacities

Specific heat capacity of agglomerated concrete: $c = 879\,J.kg^{-1}$

Specific heat capacity of stone: $c = 881\,J.kg^{-1}$

Specific heat capacity of steel: $c = 465\,J.kg^{-1}.K^{-1}$

Specific heat capacity of air: $c_V = 716\,J.kg^{-1}.K^{-1}$

Thermal conductibilities

Thermal conductibility of glass wool: $\lambda = 0.045 W.m^{-1}.K^{-1}$

Thermal conductibility of rock wool: $\lambda = 0.045 W.m^{-1}.K^{-1}$

Thermal conductibility of agglomerated concrete: $\lambda = 1.4 W.m^{-1}.K^{-1}$

Thermal conductibility of stone: $\lambda = 3 W.m^{-1}.K^{-1}$

Thermal conductibility of steel: $\lambda = 54 W.m^{-1}.K^{-1}$

Thermal conductibility of glass: $\lambda = 1.05 W.m^{-1}.K^{-1}$

Thermal conductibility of air: $\lambda = 0.026 W.m^{-1}.K^{-1}$

Fusion heat of ice: $L_f = 333.55 kJ.kg^{-1}$

EXAMPLE 4.8.– 1812. Retreat from Russia where "war is always cold"!

One month after defeating Kutuzov's Russian army in Moscow, in September, Napoleon decided to leave Moscow and bring the "Great Army" back to Europe. This "retreat" was a disaster due to the harsh winter conditions and decimated almost all of the imperial army. Today's problems will be looked at in this context.

Episode 1. In the isba

A group of retreating grenadiers came across an isba that was still inhabited. After chasing away its Russian occupants (moujiks), throwing them out into the cold and misery, they settled into the isba. They discovered wood and a good number of bottles of Vodka.

Let us recall that an **isba** (in Russian: изба) is a Russian house constructed from wood, which was the traditional dwelling of Russian countrymen for a long time.

The isba is heated by a wood fire that maintains a homogeneous temperature $T_{i0} = 12\,°C$, which is very pleasant given that the temperature outside is $T_e = -30°C$. The grenadiers therefore decide to stay in this dwelling as long as the temperature is maintained.

The walls and the roof of an isba are built from wooden logs. The walls and roof here will be assimilated to homogeneous walls with a thickness of $e = 15\,cm$. The geometry and dimensions of this construction are given in the figure below. From the point of view of thermal losses, we will assimilate the doors and windows to walls. We will consider that no flow will escape through the ground.

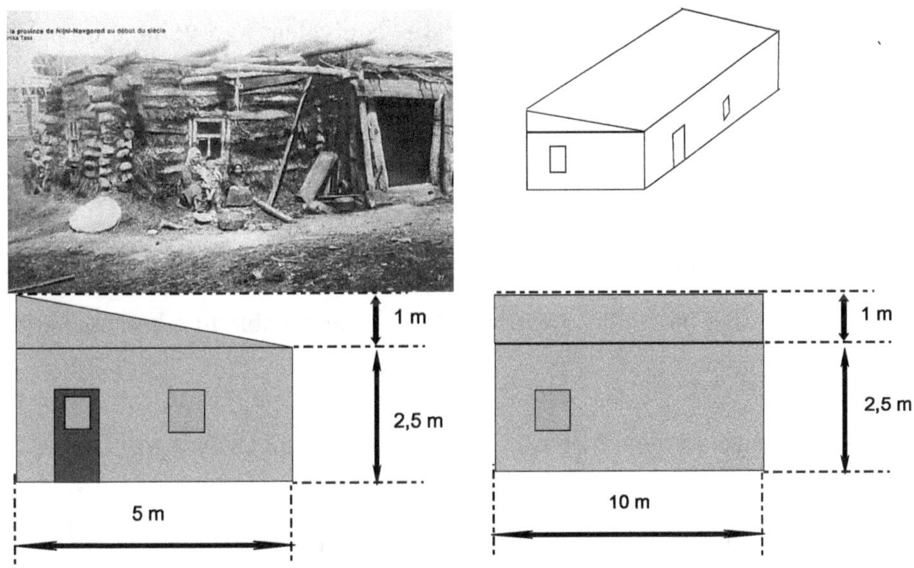

Figure 4.3. *An example of an isba and a model of it for the problem. For a color version of this figure, see www.iste.co.uk/ledoux/heat1.zip*

The density of wood for heating is $\rho = 675\,k.m^{-3}$

The combustion of one kilogram of heating wood provides $Q = 14\,MJ$.

The thermal conductibility of wooden logs is $\lambda = 0.15\,W.m^{-1}.K^{-1}$.

We will take into account an internal convection coefficient $h_i = 5 W.m^{-1}.K^{-1}$ and an external coefficient resulting from a strong wind $h_e = 30 W.m^{-2}.K^{-1}$.

1) Calculate the flow Φ of thermal losses through the walls and the roof of the isba. Deduce from this, the mass of wood m that must be burned each hour to maintain the temperature in the room.

2) The grenadiers have discovered a stock of two steres of wood in a storage area. How long will they be able to stay in the isba before continuing their retreat?

We should recall that a stere of wood is equivalent to a volume of wood of 1m^3.

SOLUTION.–

1) We must first calculate the overall thermal resistance R_{th} of the wall.

The wall is simply made of thick wood. The thermal resistance of each layer is calculated using the equation: $R_{th} = \dfrac{e}{\lambda}$, applicable to a homogeneous plane wall.

The resistance to the transfer of each convection coefficient will be:

$$R_{th\,conv} = \frac{1}{h} \qquad [4.174]$$

$$e = 15.10^{-2}\, m$$

$$\lambda = 0.15\; W.m^{-2}.K^{-1}$$

$$R_{mur} = \frac{e}{\lambda} = 1\, K.m^2.W^{-1} \qquad [4.175]$$

Using additivity of resistances to transfer, the resulting thermal resistance is therefore:

$$R_{th} = \frac{e}{\lambda} + \frac{1}{h_i} + \frac{1}{h_e} = \frac{15.10^{-2}}{0.15} + \frac{1}{5} + \frac{1}{30} = 1.233\, K.m^2.W^{-1} \qquad [4.176]$$

The outgoing flow of heat for a surface area S of the isba will be calculated by:

$$\Phi = S_{Isba}\,\varphi = S\frac{T_i - T_e}{R_{th}} \qquad [4.177]$$

$$\varphi = \frac{T_i - T_e}{R_{th}} \qquad [4.178]$$

The surface of walls, to which we add the ceiling:

Surface of the facade wall is:

$$S_{fac} = 2.5 * 10 = 25\, m^2 \qquad [4.179]$$

Surface of the background wall is of the form:

$$S_{back.wall} = 3.5 * 10 = 35\, m^2 \qquad [4.180]$$

Surface of a lateral wall is of the form:

$$S_{lat} = (5 * 2.5) + \left(\frac{5 * 1}{2}\right) = 12.5 + 2.5 = 15\, m^2 \qquad [4.181]$$

Width of the roof: $l = \sqrt{5^2 + 1^2} = 5.1\, m^2$ [4.182]

Area of the roof: $S_{roof} = 5.1 * 10 = 51\, m^2$ [4.183]

Total surface area of the isba:

$$S_{Isba} = S_{fac} + S_{fond} + (2 * S_{lat}) + S_{roof} \qquad [4.184]$$

$$S_{Isba} = 25 + 35 + (2*15) + 51 = 141 m^2 \quad [4.185]$$

$$\varphi = \frac{12-(-30)}{1.233} = 34.06 W.m^{-2} \quad [4.186]$$

$$\Phi = S\phi = 141*34.06 = 4802.5 W \quad [4.187]$$

Each hour, the following energy is required

$$E = 4802.5*3600 W = 1.73.10^7 J \quad [4.188]$$

to compensate for losses.

This energy will be given out by the hourly combustion of a mass m of wood:

$$m = \frac{1.73.10^7}{14.10^6} = 1.23 kg.hr^{-1} \quad [4.189]$$

2) The grenadiers have two steres of wood ($V_{OL} = 2m^3$), in other words a mass

$$M = \rho V_{OL} = 675*2 = 1350 kg \quad [4.190]$$

They will therefore have

$$t_S = \frac{1350}{1.23} = 1098 \; h \quad [4.191]$$

in other words:

$$t_S = 45.75 \; days \quad [4.192]$$

Episode 2. A strong Vodka

One of the grenadiers discovered a full bottle of Vodka, and he puts it in his bag to warm himself up when necessary.

He goes out of the isba. The outside temperature is as it was before, $T_e = -30°C$. The air in the bag is at the same temperature.

At this moment, the temperature of the Vodka in the bottle is $T_{i0} = 12°C$.

The Vodka, which contains 60% water is likely to freeze. Its volume then increases, and the full bottle that contains it explodes.

The Vodka freezes at $T_F = -17°C$.

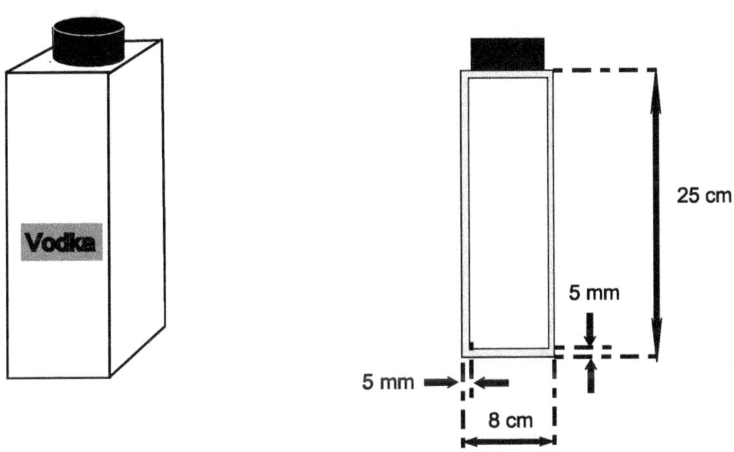

Figure 4.4. *The bottle of Vodka and its model. For a color version of this figure, see www.iste.co.uk/ledoux/heat1.zip*

The bottle is modeled by a parallelepiped of glass with a thickness of $e = 5\,mm$ and internal dimensions of $25 \text{ cm} * 8 \text{ cm} * 8 \text{ cm}$. We will only take into account the conduction on all four sides and the bottom of the bottle[1].

The conductibility of glass is $\lambda_V = 1.05 W.m^{-1}.K^{-1}$, the volumetric mass of the Vodka is $\rho = 916\,kg.m^{-3}$ and its specific heat capacity is $c = 3{,}500 J.kg^{-1}$.

1 We therefore neglect the losses through the neck and the cork here.

We will consider that the external wall of the bottle is at the external atmospheric temperature.

1) We will consider that the Vodka is then cooled by a quasi-stationary process.

a) Write the relationship between the instantaneous flow of heat $\Phi(t)$ that enters the bottle and the instantaneous internal and external temperatures $T_i(t)$ and T_e respectively.

b) Deduce from this, the differential equation for $T_i(t)$.

2) Solving the differential equation. How long after the grenadier goes out will the bottle explode?

3) Is the Vodka completely lost?

4) One of the grenadiers did not follow his companions.

He has found a hidden stock of 50 bottles of Vodka that are identical to those carried by the other grenadiers. He says to himself that by emptying them onto the fire, he will be able to stay a little longer in the warm hut. He does this, emptying the bottles little by little to maintain the internal temperature, constant at $T_{i0} = 12°C$.

The combustion of a kilogram of Vodka provides $Q' = 11.9\,MJ.kg^{-1}$.

How much longer will the grenadier who has stayed behind be able to stay in the isba? Is his idea a good one?

SOLUTION.–

1) We presume a quasi-stationary process

a) We calculate the thermal resistance R_{th} of the walls of the bottle, via the formula that is applicable to a homogeneous plane wall.

$$e = 5.10^{-3}\,m$$

$$\lambda = 1.05\,W.m^{-2}.K^{-1}$$

$$R_{glass} = \frac{5.10^{-3}}{1.05} = 4.76.10^{-3}\ K.m^2.W^{-1} \quad [4.193]$$

The internal surface of the bottle through which losses pass is composed:

of four sides with area $S_C = 8.10^{-2} * 25.10^{-2} = 2.10^{-2}\, m^2$ [4.194]

of a base with surface area $S_F = 8.10^{-2} * 8.10^{-2} = 6.4.10^{-3}\, m^2$ [4.195]

In other words, the total exchange area:

$$S = 4S_C + S_F = 8.64.10^{-2}\, m^2 . \quad [4.196]$$

Presuming a stationary process, the flow of heat Φ that exits the bottle for a value $T_i(t)$, for an area $S = 40\, m^2$ is calculated by:

$$\varphi = S\varphi = S\frac{T_e - T_i}{R_{th}} \quad [4.197]$$

b) For a time interval dt, the flow of heat exiting the bottle Φdt leads to heating dT_i such that:

$$\Phi\, dt = -\rho c V_{OL}\, dT_i \quad [4.198]$$

The minus sign expresses that the outgoing heat leads to the cooling of the Vodka.

Replacing Φ by its expression, introducing the variable θ, we find the differential equation:

$$\theta = T_i(t) - T_e \quad [4.199]$$

$$\rho_{air}\, c V_{OL} \frac{dT_i}{dt} = -S\frac{\theta}{R_{th}} \quad [4.200]$$

which can be rewritten, noting that:

$$\frac{dT_i}{dt}=-\frac{d(T_i-T_e)}{dt}=-\frac{d\theta}{dt} \qquad [4.201]$$

$$\frac{d\theta}{dt}=-\alpha\theta \qquad [4.202]$$

with

$$\alpha=\frac{S}{\rho c V_{OL} R_{th}} \qquad [4.203]$$

The volume of liquid within the bottle is

$$V_{OL}=8.10^{-2}*8.10^{-2}*25.10^{-2}=1.6.10^{-3}m^2=1.6l \qquad [4.204]$$

α gives

$$\alpha=\frac{8.64.10^{-2}}{916*3500*1.6.10^{-3}*4.76.10^{-3}}=3.54.10^{-3}s^{-1} \qquad [4.205]$$

This equation is associated with the limit condition:

At the initial moment in time

$$t=0 \quad ; \quad \theta=\theta_0=T_{\text{int}0}-T_e=12-(-30)=42°C \qquad [4.206]$$

2) This equation can be resolved classically

$$\frac{d\theta}{dt}=-\alpha\theta \qquad [4.207]$$

$$\frac{d\theta}{\theta}=-\alpha dt \qquad [4.208]$$

$$d\,Ln\theta=-\alpha t+LnC \qquad [4.209]$$

C is a constant. The choice of a constant in LnC means we can immediately write:

$$\theta(t) = C\exp{-\alpha t} \tag{4.10}$$

The condition at the initial time immediately gives $C = \theta_0$ [4.211]

$$\theta(t) = \theta_0 \exp{-\alpha t} \tag{4.212}$$

The bottle will explode when $T_i = -17°C$,

that is $\theta = \theta_G = -17 - (-30) = 13°C$ [4.213]

This will occur at a time t_G such that:

$$\theta_G = \theta_0 \exp{-\alpha t_G} \tag{4.214}$$

In other words:

$$t_G = \frac{-1}{\alpha} Ln \frac{\theta_G}{\theta_0} \tag{4.215}$$

$$t_G = \frac{-1}{3{,}538.10^{-3}} Ln \frac{13}{42} \tag{4.216}$$

$$t_G = 331{,}5\,s = 5\,mn\ 31\,s \tag{4.217}$$

3) As long as it is frozen, the Vodka remains in a block and can be recovered. All we need to do is make it melt in a bowl. However, we need a bowl and some fire! We could also put it in a goatskin and place it in the stomach of a horse that has recently died.

4) Combustion of the 50 bottles provides an energy E_V equal to:

$$E_V = 50\,\rho V_{OL}\, Q' \tag{4.218}$$

$$E_V = 50 * 916 * 1.6.10^{-3} * 11.9.10^{6} = 8.7.10^{8}\ J \tag{4.219}$$

4802.5 J is needed to compensate for the losses for a second.

The grenadier who stayed behind will have an extra time of

$$t_L = \frac{8.7.10^8}{4802.5} = 1.81.10^5\, s \qquad [4.220]$$

$$t_L = 50\,h\ 26\,mn \qquad [4.221]$$

This is all right. He could have kept one aside to warm himself up in the snow.

4.5. Applications for axisymmetric systems

EXAMPLE 4.9.– Steel bar and glass sleeve.

Part one. Static study

A steel bar with a diameter $d_i = 2\,r_i = 5\,cm$ is surrounded by a glass sleeve with an external diameter $d_{e1} = 2\,r_{e1} = 12\,cm$. On the outside of the sleeve, the convection coefficient is $h = 8\,Wm^{-2}K^{-1}$. The thermal conductibility of the glass is $\lambda = 1.05\,Wm^{-1}K^{-1}$

The tube is maintained at a temperature $T_B = 70°C$. The external temperature is $T_e = 21°C$.

1) Calculate the flow of heat between the tube and the external atmosphere for a tube length of 1 meter.

2) We replace the glass sleeve of external diameter 12 cm by a new glass sleeve with external diameter $d_{e2} = 2\,r_{e2} = 26.3\,cm$. What happens to the heat flow?

3) We replace the glass sleeve of external diameter 23.6 cm by a new glass sleeve with external diameter $d_{e3} = 2\,r_{e3} = 36\,cm$. What happens to the heat flow?

4) Compare the flows found in questions 1–3. What do you observe? How do you explain this result?

For use if needed, we give:

Density of air: $\rho_c = 1.3\, kg.m^{-3}$ [4.222a]

Specific heat capacity of air, at constant volume:

$$c_V = 716\, J.kg^{-1} \quad [4.222b]$$

Density of steel: $\rho_{steel} = 7{,}833\, kg.m^{-3}$ [4.222c]

Specific heat capacity of steel: $c_{steel} = 465\, J.kg^{-1}$ [4.222d]

Thermal conductibility of steel: $\lambda_{steel} = 54\, W.m^{-1}.K^{-1}$ [4.222e]

Thermal conductibility of glass: $\lambda_{glass} = 1.05\, W.m^{-1}.K^{-1}$ [4.221f]

SOLUTION.–

1) We must calculate the thermal resistance of the sleeve (which will be per meter of length), then the thermal flow.

The thermal resistance is given by adding the thermal resistance of the glass sleeve and the thermal resistance due to convection:

$$R_{th} = \frac{Ln\frac{r_{e1}}{r_i}}{2\pi\lambda} + \frac{1}{2\pi h} = \frac{Ln\frac{d_{e1}}{d_{i1}}}{2\pi\lambda_S} + \frac{1}{2\pi h} \quad [4.223]$$

We obtain:

$$R_{th/m} = 0.464\, KmW^{-1} \quad [4.224]$$

The difference in temperature between the two sides of the sleeve is:

$$\Delta T = T_B - T_e = 70 - 21 = 49°C \quad [4.225]$$

In other words, the thermal flow evacuated to the exterior per meter of bar:

$$\Phi = L\frac{\Delta T}{R_{th1}} \qquad [4.226]$$

with $L = 1\,m$ [4.227]

$$\Phi = \frac{49}{0.464} = 105.6\,W \qquad [4.228]$$

2) We continue the same calculation with the new value of external diameter $d_{e2} = 2r_{e2} = 26.3\,cm$. We note that physically, this value has an influence on the conductive transfer in the glass, as well as on the convection to the outside.

We then obtain

$$R_{th/m} = 0.4029\,K\,mW^{-1} \qquad [4.229]$$

$$\Phi = 121.6\,W \qquad [4.230]$$

3) Again, we continue by applying the same calculation with the new value of the external diameter $d_{e2} = 2r_{e2} = 26.3\,cm$. We note that physically, this value influences the conductive transfer in the glass, as well as the convection to the outside.

We also find

$$R_{th/m} = 0.4097\,K\,mW^{-1} \qquad [4.231]$$

$$\Phi = 119.6\,W \qquad [4.232]$$

4) Here, we must think of the phenomenon of "**critical radius**".

We see that:

– from 12 cm to 26.3 cm, the diameter increases and the losses increase;

– from 26.3 cm to 36 cm, the diameter increases and the losses reduce.

We know that when we increase the thickness of the insulating sleeve, there is competition between the increase in insulation, which tends to reduce losses, and the increase in the external lateral surface area of the entire object, which tends to increase losses. When the external radius **is less than a "critical" radius**, the competition is to the advantage of the external surface area and if the thickness of insulation in this zone is increased, this corresponds paradoxically to an increase in losses. When the **external radius is higher than the critical radius**, the trend is inverted and increasing the thickness of the insulation then contributes to reducing losses.

In this case, the value of the critical radius is:

$$r_C = \frac{\lambda}{h} = \frac{54}{8} = 0.131\, m = 13.1\, cm \qquad [4.233]$$

At $d_{e2} = 2r_{e2} = 26.3\, cm$ of diameter, we are practically at the critical radius

Therefore:

– for $d_e < r_c = 26.2\, cm$ (from $12\,\text{cm}$ to $26.3\,\text{cm}$), the diameter increases and the losses increase;

– for $d_e > r_c = 26.2\, cm$ (from $26.3\,\text{cm}$ to $36\,\text{cm}$), the diameter increases and the losses are reduced.

Part two. Steel bar and glass sleeve; evolution over time

We will look again at the steel bar from the previous example, surrounded by the glass sleeve with diameter $d_{e2} = 2r_{e2} = 26.3\, cm$. The external convection coefficient is still $h = 8\, Wm^{-2}K^{-1}$.

In addition, we give:

Volumetric mass of steel: $\rho_S = 7{,}833\, kg\, m^{-3}$

Specific heat capacity of steel: $c_S = 465\, J\, kg^{-1}K^{-1}$

In fact, the thermal losses lead to a reduction in the temperature of the bar, in line with a law $T_B(t)$ that will be calculated. To do so, we will use a quasi-stationary balance calculation.

We will neglect the thermal losses through the extremities of the bar and the glass sleeve.

1) Write the relationship between the flow exiting the bar per meter of length, Φ, $T_B(t)$ and T_e

2) Deduce from this, the differential equation that connects $\frac{dT_B}{dt}$ to T_B and T_e.

Then, deduce the equation for $\frac{dT_B}{dt}$.

3) Give the expression for $T_B(t)$.

4) After a certain time period $t_{0,5}$, will the temperature of the bar have reduced by half? Give this time in hours, minutes and seconds.

We recall, to be used as needed:

Density of air: $\rho_c = 1.3\, kg.m^{-3}$

Specific heat capacity of air, of constant volume: $c_V = 716\, J.kg^{-1}$

Density of steel $\rho_{steel} = 7{,}833\, kg.m^{-3}$

Specific heat capacity of steel: $c_{steel} = 465\, J.kg^{-1}$

Thermal conductibility of steel: $\lambda_{steel} = 54\, W.m^{-1}.K^{-1}$

Thermal conductibility of glass: $\lambda_{glass} = 1.05\, W.m^{-1}.K^{-1}$

SOLUTION.–

1) By definition of the thermal resistance:

$$\Phi = \frac{T_B - T_e}{R_{th}} \qquad [4.234]$$

We had calculated:

$$R_{th/m} = 0.4029\, K\, m W^{-1} \qquad [4.235]$$

2) A thermal balance for a meter of bar gives us the differential equation, which is satisfied by T_B:

$$\rho_S c \frac{dT_B}{dt} = -\frac{T_B - T_e}{R_{th}} \qquad [4.236]$$

with the initial condition:

$$t = 0 \quad ; \quad T_B = T_{Bi} = 70°C \qquad [4.237]$$

The sign "$-$" in the equation indicates that the flow of heat (positive) contributes to the reduction in the temperature; the derivative is therefore a negative term.

3) Resolution of this equation is classic:

The variable $\theta = T_B - T$ is introduced [4.238]

The equation becomes:

$$\rho_S c\, V_{OL} \frac{d\theta}{dt} = \frac{\theta}{R_{th}} \qquad [4.239]$$

V_{OL} is here the volume of a meter of bar, in other words:

$$V_{OL} = \pi r_i^2 = \pi \left(2.5.10^{-2}\right)^2 = 1.96.10^{-3}\, m^3 \qquad [4.240]$$

or:

$$\frac{d\theta}{\theta} = -\frac{dt}{\rho_S c R_{th}} = -\alpha\, dt \qquad [4.241]$$

with $\alpha = \dfrac{1}{\rho_S c V_{OL} R_{th}}$ [4.242]

with the initial condition that becomes:

$$t = 0 \quad ; \quad \theta = T_{Bi} - T_e = 49°C \qquad [4.243]$$

Integrating the two terms of the equation:

$$Ln\,\theta = -\alpha t + Ln C \qquad [4.244]$$

A constant LnC appears, which will be given by the initial condition. The choice of the constant in log form is a trick that allows us to simplify the change to an exponential for the preceding expression. Taking the exponential for each term:

$$\exp(Ln\,\theta) = \exp(-\alpha t)\exp(LnC) \qquad [4.245]$$

$$\theta = C\exp(-\alpha t) \qquad [4.246]$$

at

$$t = 0 \qquad T_{Bi} - T_e = C = 49 \qquad [4.247]$$

Hence, the final expression for the temperature of the bar:

$$T_B(t) - T = (T_{Bi} - T)\exp(-\alpha t) \qquad [4.248]$$

4) We calculate

$$\alpha = \frac{1}{\rho c V_{OL} R_{th}} = \frac{1}{7833 * 465 * 1.96.10^{-3} * 0.4029} = 3.47.10^{-4} \qquad [4.249]$$

We want to change the temperature from $T_B = 70°C$ to $T_B = 35°C$

in other words, $T_{Bi} - T$ from $T_{Bi} - T = 49°C$ to $T_{Bi} - T = 35 - 21 = 14°C$

The time required, t_S, corresponds to the relationship:

$$14 = 49\exp(-\alpha t_S) \quad [4.250]$$

or even:

$$14 = 49\exp(-\alpha t_S) \quad [4.251]$$

$$t_S = \frac{-1}{\alpha} Ln\frac{14}{49} = \frac{1}{3.47.10^{-4}} Ln\frac{49}{14} \quad [4.252]$$

$$t_S = 3610\,s = 1\,hr\,10\,s \quad [4.253]$$

EXAMPLE 4.10.– Insulation of a laboratory chemical reactor

A chemist wishes to maintain a mixture containing $m = 15\,kg$ of reagents for a **minimum** time of **one half-hour** at a constant temperature, to the nearest **one and a half degrees**.

To do so, he places his reagents into a pyrex cylinder with an internal diameter d_i and external diameter d_e. The tube is horizontal and of length L. The tube is fixed by thick supports on its right-hand and left-hand faces. Throughout the problem, we will neglect the transfers through these two lateral faces.

In the calculations, we will neglect flows of quantities of heat that are required for the variations in temperature of the pyrex.

The atmospheric temperature in the laboratory is set by the administration to $T_e = 19°C$. The initial temperature of the reagents is $T_{ri} = 100°C$.

We will apply an external convection coefficient h.

1) The chemist is not a great physicist; he begins his experiment with the simple glass tube.

a) Calculate the thermal resistance per meter of the glass cylinder, R_{th1}, and deduce from this the flow Φ_{verre1} that escapes from the reagents at the initial instant of the reaction.

b) Give an approximate estimation of the reduction in temperature of the reagents in a second right at the beginning of the experiment.

Do you intuitively believe, without any other calculation, that the temperature can be maintained constant to the nearest degree for two hours?

2) On the basis of good advice from one of their physicist colleagues, they seek to improve the process by encasing its reactor in a rock wool sheath of thickness e.

We have $T(t)$ as the temperature of the products during cooling.

a) Give the expression[2] for the resultant thermal resistance R_{th2} for this new reactor. Calculate its numerical value.

b) Establish that, taking into account this new value R_{th2}, the flow $\Phi_{isol}(t)$ exiting this new reactor configuration can be written as $\Phi = \beta(T_i - T_e)$; β is a coefficient whose numerical value will be given.

c) Then, write a differential equation for $T(t)$.

d) Give the expression for $T(t)$. To do so, it will be useful to define a variable $\theta(t) = T(t) - T_e$

e) By how much will the temperature of the reagents reduce after half an hour? Who is better, the physicist or the chemist?

2 Since the thickness is not known, this question will be treated literally up to **e**.

Numerical data:

$d_i = 5\,cm$

$d_e = 6\,cm$

$e = 3\,cm$

$L = 1.5\,m$

$T_{ri} = 100°C$

$T_e = 19\,C$

$h = 5\,W.m^{-2}.K^{-1}$

$m = 15\,kg$

Specific heat capacity of the reagents: $c = 4{,}000\,J.kg^{-1}$

Thermal conductibility of the rock wool: $\lambda_{LR} = 0.045\,W.m^{-1}.kg^{-1}$

Thermal conductibility of pyrex: $\lambda_{Pyr} = 1.13\,W.m^{-1}.kg^{-1}$

SOLUTION.–

1) We begin the experiment with the simple glass tube

a) The thermal resistance R_{thLR} *per meter* of a cylindrical sleeve with internal radius r_1, external radius r_2 and thermal conductibility λ is calculated from the expression:

$$R_{th} = \frac{Ln\left(\frac{r_e}{r_i}\right)}{2\,\pi\,\lambda} \qquad [4.254]$$

to which the resistance to the transfer due to convection is added

$$R_{conv} = \frac{1}{2\pi r_e \lambda} \quad [4.255]$$

R_{th1} is calculated using the equation drawn up in the example

$$R_{th1} = \frac{Ln\frac{r_e}{r_i}}{2\pi\lambda} + \frac{1}{2\pi r_e h} = \frac{Ln\frac{6}{5}}{2\pi 3} + \frac{1}{2\pi 3.10^{-2}.5}$$

$$= 2.5679.10^{-2} + 1.061 = 1.0867 W m^{-1} K^{-1} \quad [4.256]$$

$$T_i - T_e = 100 - 19 = 81° \quad [4.257]$$

$$\Phi = L\frac{T_i - T_e}{R_{th1}} = 1.5.\frac{81}{1.0867} = 111.81W \quad [4.258]$$

b) To reduce the temperature of the reagents by one degree:

$$Q = mc = 15.4000 = 60,000 J \quad [4.259]$$

$$\Delta t = 1s \ ; \ \Delta T \approx \frac{111.81}{60000} = 1.8610^{-3°} \quad [4.260]$$

1.5 degrees over half an hour represents "on average" $8.33.10^{-4}$ °s^{-1}. Significant doubts remain over the order of magnitude. A more comprehensive calculation is required.

2) We are attempting to improve the process by encasing its reactor in a rock wool sheath of thickness e.

a) We will calculate the thermal resistance R_{th2} that results from this new reactor.

$$R_{th1} = \frac{Ln\frac{r_e}{r_i}}{2\pi\lambda_{pyrex}} + \frac{Ln\frac{r_e+e}{r_e}}{2\pi\lambda_{LR}} + \frac{1}{2\pi r_e' h} = \frac{Ln\frac{3}{2.5}}{2\pi * 1.13} + \frac{Ln\frac{6}{3}}{2\pi * 0.045} + \frac{1}{2\pi * 6.10^{-2} * 5}$$

$$= 2.5679.10^{-2} + 2.451 + 0.5305 = 3.0077 \quad [4.261]$$

$$r_e^{'} = 6.10^{-2}\, m \qquad [4.262]$$

b) The flow $\Phi_{isol}(t)$ exiting this new configuration of reactor can be written as:

$$\Phi = \beta(T_i - T_e) \qquad [4.263]$$

$$\Phi = L\frac{T_i - T_e}{R_{th1}} = 1.5.\frac{T_i - T_e}{3.0077} = 0.499(T_i - T_e) \qquad [4.264]$$

Deducing from this: [4.265]

$$\beta = 0.499 \qquad [4.266]$$

c) Heat balance: by writing that the heat exiting per unit of time cools the reagents, we draw up the differential equation for $T(t)$

$$m\, c\, dT_i = -\Phi \qquad [4.267]$$

$$\Phi = \beta\,(T_i - T_e) \qquad [4.268]$$

with

$$\beta = 0.499 \qquad [4.269]$$

$$mc\frac{dT_i}{dt} = -\beta\,(T_i - T_e) \qquad [4.270]$$

d) We will introduce the variable $\theta = (T_i - T_e)$ [4.271]

$$\Phi = \beta\,(T_i - T_e) = \beta\,\theta \qquad [4.272]$$

$$\beta = 0.499 \qquad [4.273]$$

$$mc\frac{dT_i}{dt}=-\Phi=-\beta\left(T_i-T_e\right) \quad [4.274a]$$

$$\theta=\left(T_i-T_e\right) \quad [4.274b]$$

$$\frac{d\theta}{dt}=-\alpha\theta \quad [4.274c]$$

$$\alpha=\frac{\beta}{mc}=\frac{0.499}{15*4000}=8.31.10^{-6} \quad [4.275]$$

$$t=0\ ;\ \theta=\theta_0=100-19=81° \quad [4.276]$$

$$\frac{d\theta}{\theta}=-\alpha dt \quad [4.277]$$

$$Ln\theta=-\alpha t+Ln\,C \quad [4.278]$$

$$\theta=C\exp-\alpha t \quad [4.279]$$

$$\theta=\theta_0\exp-\alpha t \quad [4.280]$$

e) By how much will the temperature of the reagents reduce after half an hour.

$$\theta=\theta_0\exp-\alpha t \quad [4.281]$$

$$\theta_0 = 81° \quad [4.282]$$

$$\alpha= 8.31.10^{-6} \quad [4.283]$$

$$t=0.5\,hr=1{,}800\,s \quad [4.284]$$

$$\theta=81\exp-8.31.10^{-6}.1{,}800=0.985.81=79.98 \quad [4.285]$$

$$\Delta T=1.2° \quad [4.286]$$

Mission accomplished. The physicist is indeed better.

EXAMPLE 4.11.– A good beer must be cold

A bottle of beer is assimilated to a glass cylinder[3] of height $H = 26\,cm$, with internal diameter $D_i = 6\,cm$ and thickness $e = 5\,mm$.

On a hot summer day, a bottle of beer has been left next to a window. The temperature of the beverage has reached $T_0 = 25°C$.

To drink a cold beer, at $T_{F1} = 6°C$, a tourist places the bottle in the main body of the fridge, which is at a temperature of $T_{e1} = 4°C$. We will consider that the natural convection imposes a uniform convection coefficient on the external surface of the bottle, $h_e = 5\,W.m^{-2}.K^{-1}$. We will also consider that the temperature $T_i(x)$ of the beer is uniform at all times in the bottle and that, as a consequence, the cooling process is quasi-stationary.

In addition, we point out that the beer freezes at $T_G = -2°C$ and that, in this case, the beer dilates and the bottle explodes.

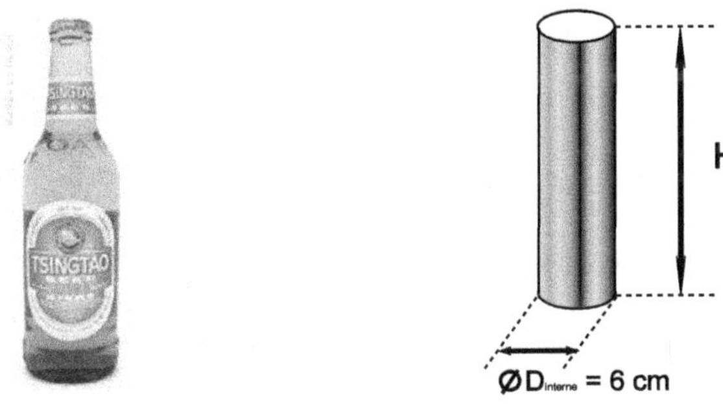

Figure 4.5. *The bottle and its model*

1) The bottle is placed in the fridge

a) Calculate the thermal resistance R_{th} per meter of length of the glass in the bottle.

3 Here, we will neglect the thermal effects of the neck and the bottle cap.

b) Calculate the thermal flow Φ_0 coming out of the bottle at the initial moment in time; in other words, when the bottle is placed in the fridge.

2) The beer cools

a) Give the differential equation that satisfies the temperature $T_i(x)$ of the beer in the bottle.

b) Give the expression for $T_i(x)$.

c) How long will the tourist have to wait before drinking their cold beer?

3) Since this time is too long, the tourist's wife, who is not really a physicist, places a second bottle of beer, also at $T_0 = 25°C$ in the freezer compartment, where the temperature is $T_{e2} = -15°C$.

She thinks that the beer should be left for a shorter period in the freezer and leaves it for a time $t_2 = 1\ hour$

a) Will she drink a very cold beer?

b) Estimate the minimum time necessary for the bottle to explode.

c) What do you think of all this?

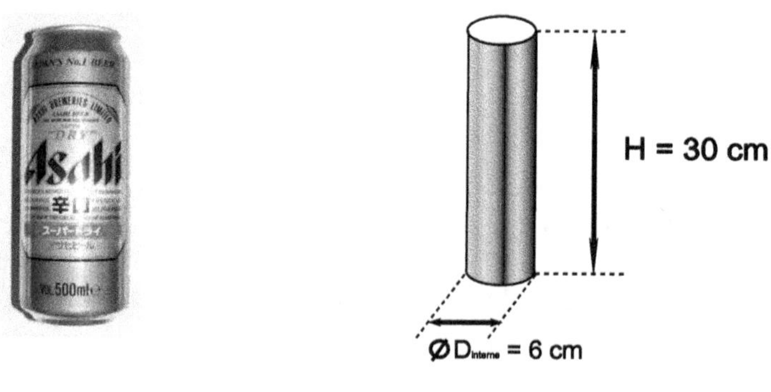

Figure 4.6. *The can and its model*

4) Looking again at questions 1 and 2 for a can of beer, it is assimilated to an aluminum cylinder with height $H = 30\,cm$, internal diameter $D_i = 6\,cm$ and thickness $e = 1\,mm$.

We will conserve the same convection coefficient h_e.

Additional data:

Volumetric mass of the beer: $\rho = 1{,}000\,kg.m^{-3}$

Specific heat capacity of the beer: $c = 4{,}180\,J.kg^{-1}$

Thermal conductibility of the beer: $\lambda = 0.6\,W.m^{-1}.K^{-1}$

Thermal conductibility of the glass: $\lambda = 1.05\,W.m^{-1}.K^{-1}$

Thermal conductibility of aluminum: $\lambda = 204\,W.m^{-1}.K^{-1}$

SOLUTION.–

1) At the initial moment in time

a) We calculate the thermal resistance R_{th} of the sides of the bottle, using the equation:

$$R_{th} = \frac{Ln\frac{r_e}{r_i}}{2\pi\lambda} = \frac{Ln\frac{r_e}{r_i}}{2\pi\lambda} \qquad [4.287]$$

which is applicable to a homogeneous cylinder.

$$r_i = 3.10^{-2}\,m$$
$$r_e = 3.5.10^{-2}\,m$$
$$D_i = 6.10^{-2}\,m$$
$$D_e = 7.10^{-2}\,m$$
$$\lambda = 1.05\,W.m^{-2}.K^{-1}$$

$$R_{th} = \frac{Ln\frac{r_e}{r_i}}{2\pi\lambda} = \frac{Ln\frac{7}{6}}{2*\pi*1.05} = 2.34.10^{-2}\,W.m^{-2}.K^{-1} \quad [4.288]$$

b) The total thermal resistance including the convection coefficient is:

$$R_{tot} = \frac{Ln\frac{r_e}{r_i}}{2\pi\lambda} + \frac{1}{2\pi r_e h_e} = 2.34.10^{-2} + \frac{1}{2*\pi*3.3.10^{-2}*20}$$

$$= 2.34.10^{-2} + 0.241 = 0.264\,W.m^{-2}.K^{-1} \quad [4.289]$$

The flow of heat Φ exiting the bottle for an initial value T_i $25°C$ is calculated for a cylinder length $H = 26\,cm$, in other words:

$$\varphi_0 = H\frac{T_i - T_e}{R_{th}} = 0.26*\frac{25-4}{0.264} = 20.7\,W \quad [4.290]$$

2) In a quasi-stationary model, the instantaneous flow exiting the bottle is written as:

$$\varphi(t) = H\frac{T_i(t) - T_e}{R_{tot}} \quad [4.291]$$

a) For a time dt , the heat flow exiting the bottle Φdt leads to heating dT_i such that:

$$\Phi\, dt = -\rho c V_{OL}\, d T_i \quad [4.292]$$

The minus sign expresses that the outgoing heat leads to the cooling of the liquid.

Replacing Φ by its expression and introducing the variable θ , we find the differential equation:

$$\theta = T_i(t) - T_e \quad [4.293]$$

$$\rho_{beer}\, c V_{OL} \frac{dT_i}{dt} = -H \frac{\theta}{R_{tot}} \quad [4.294]$$

This can be rewritten, noting that:

$$\frac{dT_i}{dt} = -\frac{d(T_i - T_e)}{dt} = -\frac{d\theta}{dt} \quad [4.295]$$

$$\frac{d\theta}{dt} = -\alpha\theta \quad [4.296]$$

with

$$\alpha = \frac{H}{\rho c V_{OL} R_{tot}} \quad [4.297]$$

The volume of liquid contained in the bottle is

$$V_{OL} = \pi * \frac{(6.10^{-2})^2}{4} * 26.10^{-2} = 7.35.10^{-4}\, m^3 \quad [4.298]$$

α gives

$$\alpha = \frac{26.10^{-2}}{1000 * 4180 * 7.35.10^{-4} * 0.264} = 3.20.10^{-4}\, s^{-1} \quad [4.299]$$

This equation is combined with the limit condition:

At the initial instant:

$$t = 0 \quad ; \quad \theta = \theta_0 = T_{i0} - T_e = 25 - 4 = 21°C \quad [4.300]$$

b) This equation can be resolved classically:

$$\frac{d\theta}{dt} = -\alpha\theta \quad [4.301]$$

$$\frac{d\theta}{\theta} = -\alpha\, dt \qquad [4.302]$$

$$d\,Ln\theta = -\alpha t + LnC \qquad [4.303]$$

C is a constant. The choice of a constant in LnC allows us to immediately write:

$$\theta(t) = C\exp -\alpha t \qquad [4.304]$$

The condition at the initial time immediately gives $C = \theta_0$ [4.305]

$$\theta(t) = \theta_0 \exp -\alpha t \qquad [4.306]$$

c) The bottle should be taken out when $T_i = T_f = 6°C$,

so $\theta = \theta_f = 6 - 4 = 2°C$ [4.307]

This will happen at a time t_1 such that:

$$\theta_f = \theta_0 \exp -\alpha t_1 \qquad [4.308]$$

In other words:

$$t_1 = \frac{-1}{\alpha} Ln\frac{\theta_f}{\theta_0} \qquad [4.309]$$

$$t_1 = \frac{-1}{3.2.10^{-4}} Ln\frac{2}{21} \qquad [4.310]$$

$$t_1 = 7,348\ s = 2\,hr\,2\,mn \qquad [4.311]$$

3) The bottle is in the freezer compartment

a) The new value of θ_0 is $\theta_0 = 25 - (-15) = 40°C$ [4.312]

The temperature of the beer after $t_2 = 1\,hr = 3600\,s$ will then be

$$\theta_f = \theta_0 \exp - \alpha t_2 \; = 40 \exp\left(-3.2.10^{-4} * 3600\right) \quad [4.313]$$

$$\theta_f = 12.64 \quad [4.314]$$

$$T_f = 12.64 + (-15) = -2.36°C \quad [4.315]$$

And no, the bottle will have exploded. Disaster!

b) We can estimate the freezing time

The bottle will explode when $\theta_f = -2-(-15) = 13°C$, in other words after

$$t_{\exp} = \frac{-1}{3.2.10^{-4}} Ln\frac{31}{40} = 3512\,s = 58.5\,mn \quad [4.316]$$

We are not far off, it is a shame. Disaster again, but only by a small margin!

c) Unlucky

4) In the case of the metal can, it is necessary to look at the problem again, by changing the value of the thermal resistance of the container.

$$r_i = 3.10^{-2}\,m$$
$$r_e = 3.1.10^{-2}\,m$$
$$D_i = 6.10^{-2}\,m$$
$$D_e = 6.2.10^{-2}\,m$$
$$\lambda = 204\,W.m^{-2}.K^{-1}$$

$$R_{th} = \frac{Ln\frac{r_e}{r_i}}{2\pi\lambda} = \frac{Ln\frac{6.2}{6}}{2*\pi*204} = 2.56.10^{-5}\,W.m^{-2}.K^{-1} \quad [4.317]$$

We see that the thermal resistance of the metal will be negligible in comparison to that of the convection coefficient

$$R_{tot} = \frac{Ln\frac{r_e}{r_i}}{2\pi\lambda} + \frac{1}{2\pi r_e h_e} = 2.56.10^{-5} + \frac{1}{2*\pi*3.1.10^{-2}*20}$$

$$= 2.56.10^{-5} + 0.257 = 0.257 W.m^{-2}.K^{-1} \quad [4.318]$$

The initial instantaneous flow will be:

$$\varphi_0 = H\frac{T_i - T_e}{R_{th}} = 0.3*\frac{25-4}{0.257} = 24.5W \quad [4.319]$$

The volume of liquid has been modified slightly:

$$V_{OL} = \pi*\frac{\left(6.10^{-2}\right)^2}{4}*0.3 = 8.48.10^{-4}\, m^3 \quad [4.320]$$

We will have a value close to the previous one:

$$\alpha = \frac{30.10^{-2}}{1000*4180*8.48.10^{-4}*0.257} = 3.29.10^{-4}\, s^{-1} \quad [4.321]$$

And the cooling time will be modified a little in the end:

$$t_1 = \frac{-1}{3.29.10^{-4}} Ln\frac{2}{21} \quad [4.322]$$

$$t_1 = 7147\, s = 1\, hr\, 59\, mn \quad [4.323]$$

COMMENT.–

The value chosen for the convection coefficient $h_e = 20W.m^{-1}.K^{-1}$ may appear a little high for pure natural convection.

We could argue that in certain fridges, the air is mixed around.

Put more simply, we admit that **this value of h_e is an "ad hoc" value**, which involves taking into account two phenomena that intervene in two areas of the box or the bottle:

– natural convection through the top;

– direct contact of the bottle with the shelf, which necessarily imposes the temperature $T_e = 4°C$ on part of the external wall of the recipient and which automatically eliminates all convection effects on this zone.

EXAMPLE 4.12.– Thermal audit of a yurt

As and when necessary, the numerical data provided at the end of this problem will be used.

A Mongol yurt is mainly made up of a wall of sheep's wool felt of thickness e, topped with a roof, which is also made of felt.

REMARKS.– As opposed to a Kyrgyz yurt, a Mongol yurt is always a permanent habitat. **Felt** is an unwoven fabric made by pressurizing and boiling natural fibers, sometimes with a chemical treatment.

A yurt will be assimilated to a cylinder of diameter D on the base and a height H . The roof will be assimilated to a plane disk.

The convective transfers in effect have an inside convection coefficient h_i and an outside convection coefficient h_e. These coefficients are identical for the vertical walls and the roof.

We will neglect the thermal losses through the ground.

To heat the yurt, the head of the family has lit a wood fire that consumes m kg of wood per hour. A kilogram of wood provides an energy Q by combustion. We largely consider that only 50% of heat produced passes through the felt by conduction. The remaining 50% exits directly with the smoke and has no incidence on the temperature in the yurt.

We consider that **when the fire is lit**, we are in a stationary regime. The internal temperature of the yurt T_i is constant. The outside temperature is T_e .

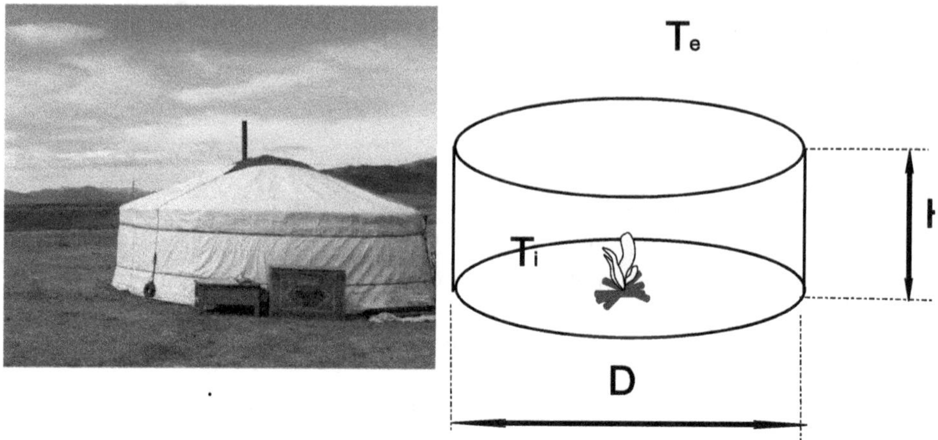

Figure 4.7. *A yurt and its heating. For a color version of this figure, see www.iste.co.uk/ledoux/heat1.zip*

1) We are attempting to evaluate the conductive thermal losses in the yurt. To do so, we will use two hypotheses.

a) In an initial model, we consider that the thermal resistance per m^2 of the lateral area of the yurt can be calculated in the same way as a plane felt wall of the same thickness (and of course of the same length as the central pillar in the yurt)[4]. The roof obviously remains a plane felt surface. Deducing from this, the flow that passes through the walls of the yurt for a temperature difference between the inside and the outside of 22°C.

b) In a second model, we take into account the cylindrical form of vertical walls. To do so, we calculate the thermal resistance of the lateral walls per meter of length. Obviously, the roof remains a plane surface. Calculate the outgoing conductive flow over the entire felt surface area of the yurt in the case where the temperature difference between the inside and the outside is 22°C.

c) Are the values found in **a** and **b** so different? Are you surprised? Which of the two calculations seems to be the most suitable to you?

---

4 This approximation is inspired by the construction of a “standard” yurt that is often carried out using five plane vertical wall panels.

2) Given the answer to the previous question, show that the correct expression for the instantaneous flow exiting the yurt as a function of the difference between the internal and external temperature takes the form

$$\Phi = \beta(T_i - T_e) \qquad [4.324]$$

where β is a coefficient whose numerical value will be given.

3) How much wood will it be necessary to burn in 24 hours to maintain the internal temperature of the yurt at 14°C?

4) With no stock left, the head of the family went to get some wood in his old Land Rover. He will take a whole day to do this. Will the family suffer from the cold?

a) Write the differential equation that is obeyed by the internal temperature $T_i(t)$.

We will maintain the relationship written in 2 when T_i (t) is variable over time (quasi-stationary calculation). The variable $\theta = (T_i - T_e)$ will be introduced to useful effect.

b) Give the literal expression for $\theta(t)$.

c) We consider that below 5°C, it is not possible to live in the yurt. How long will it take for this temperature to be reached?

Will the head of the family come back in time to keep the family comfortable?

5) Thermal science of sport.

This question can be treated independently from question 4.

We consider that a fit person pedaling vigorously on a home-trainer gives off heat of power P.

A Mongolian sports club rented a yurt that is identical to the one seen in previous questions and installed 10 home-trainers there. The club does not have the means to heat the yurt and counts on the heat given off by the cyclists to maintain an acceptable temperature in the tent.

REMARK.–

There is now no fire, so no losses via smoke going up the "chimney".

The outside temperature is $T_e = 5°C$.

a) What is the inside temperature during the stationary regime when 10 Mongol cyclists pedal inside the yurt?

b) Is it a good idea to not have installed heating?

Additional data:

$D = 6m$

$H = 2.3m$

$e = 10mm$

$T_e = -8°C$

Thermal conductibility of the felt: $\lambda_F = 0.035 W.m^{-1}.kg^{-1}$

$h_i = 5W.m^{-2}.K^{-1}$

$h_e = 16W.m^{-2}.K^{-1}$

Volumetric mass of air: $\rho = 1.3 kg.m^{-3}$

Specific heat capacity of the air: $c_v = 718 J.kg^{-1}.K^{-1}$

$Q = 14 MJ.kg^{-1}$

Heating per cyclist: $P = 400W$

SOLUTION.–

1) We are seeking to evaluate the conductive thermal losses in the yurt, looking at two models: approximation of the plane wall, cylindrical (sleeve) wall.

a) Plane wall

The thermal resistance per m² of the walls of the yurt is:

$$R_{th} = \frac{e}{\lambda} + \frac{1}{h_i} + \frac{1}{h_e} = \frac{10^{-2}}{0.035} + \frac{1}{5} + \frac{1}{16} = 0.548\ W^{-1}\,m^2{}^\circ K \quad [4.325]$$

$$S = \frac{\pi D^2}{4} + \pi D H = 28.27 + 43.35 = 71.62\,m^2 \quad [4.326]$$

$$\Phi = S\frac{\Delta T}{R_{th}} = 71.62\frac{22}{0.548} = 2875W \quad [4.327]$$

b) Cylindrical wall

The thermal resistance per meter of length of the cylinder is:

$$R_{th} = \frac{Ln\left(\frac{D_e}{D_i}\right)}{2\pi\lambda_F} + \frac{1}{2\pi r_i h_i} + \frac{1}{2\pi r_e h_e} = \frac{Ln\left(\frac{6.02}{6}\right)}{2\pi * 0.035} + \frac{1}{2\pi * 3 * 5} + \frac{1}{2\pi * 3.01 * 16}$$

$$= 1.5110^{-2} + 1.0610^{-2} + 3.3110^{-3} = 2.9.10^{-2}\,W^{-1}\,m^2{}^\circ K \quad [4.328]$$

The flow will therefore be as follows, summing up what passes through the cylinder and what passes through the circle (roof)

$$S_{Roof} = \frac{\pi D^2}{4} = 28.27\ m^2 \quad [4.329]$$

$$\Delta T = 22^\circ \quad [4.330]$$

$$\Phi = 28.27\frac{\Delta T}{0.548} + 2.3\frac{\Delta T}{2.9.10^{-2}} = 1134.9 + 1744.8 = 2897.9\ W \quad [4.331]$$

c) Are the values found in **a** and **b** close?

The thickness e is small in comparison to the radius of the yurt. We can demonstrate that the two forms of the flow are equivalent if

$$r_i \approx r_e \quad ; \quad Ln\frac{r_e}{r_i} = Ln\left(1+\frac{e}{r_i}\right) \approx \frac{e}{r_i} \qquad [4.332]$$

The form that takes account of the cylindrical geometry is more suitable.

2) The form that takes account of the cylindrical geometry is more suitable.

Therefore:

$$\Phi = 28.27\frac{\Delta T}{0.548} + 2.3\frac{\Delta T}{2.9.10^{-2}} = \beta\left(T_i - T_e\right) \qquad [4.333]$$

$$\beta = 130.9 \qquad [4.334]$$

3) What is required to maintain the internal temperature of the yurt at 14°C?

1 kg of burned wood provides 4.10^6 J.

Through conduction, in 24 hours, we lose:

$$2898 * 3600 * 24 = 2.504.10^8 \text{ J}. \qquad [4.335]$$

In total, we therefore lose (conduction + evacuation of the smoke) double this, in other words $5.077.10^8$ J.

It is therefore necessary to burn: $\dfrac{5.077.10^8}{14.10^6} = 35.77\,\text{kg}$ of wood every 24 hours.

4) The family remains without fire for an entire day.

a) We will write the differential equation that is obeyed by the inside temperature $T_i(t)$.

We define $\theta = (T_i - T_e)$ [4.336]

$$\Phi = \beta\,(T_i - T_e) \quad [4.337]$$

$$\beta = 130.9 \quad [4.338]$$

$$Vol = \pi\frac{D^2}{4}H = 65.03m^3 \quad [4.339]$$

$$\rho V_{ol}\,c\frac{dT_i}{dt} = -\Phi = -\beta\,(T_i - T_e) \quad [4.340]$$

$$\theta = (T_i - T_e) \quad [4.341]$$

$$\frac{d\theta}{dt} = -\alpha\theta \quad [4.342]$$

$$\alpha = \frac{\beta}{\rho V_{ol}\,c} = 2.156.10^{-3} \quad [4.343]$$

b) That is the literal expression for $\theta(t)$ obtained by resolving the differential equation.

$$\alpha = \frac{\beta}{\rho V_{ol}\,c} = 2.156.10^{-3} \quad [4.344]$$

$$\frac{d\theta}{dt} = -\alpha\theta \quad [4.345]$$

$$t = 0 \ ; \ \theta = \theta_0 = 14 - (-8) = 22°C \qquad [4.346]$$

$$\frac{d\theta}{\theta} = -\alpha\, dt \qquad [4.347]$$

$$Ln\,\theta = -\alpha t + Ln\, C \qquad [4.348]$$

$$\theta = C \exp - \alpha t \qquad [4.349]$$

$$\theta = \theta_0 \exp - \alpha t \qquad [4.350]$$

c) Below 5°C, it becomes impossible to live in the yurt.

If $T_i = 5C$, then $\theta = 13$ [4.351]

This intervenes at a time t such that:

$$\alpha = \frac{\beta}{\rho V_{ol}\, c} = 2.156.10^{-3} \qquad [4.352]$$

$$13 = 22 \exp - \alpha t \qquad [4.353]$$

$$t = \frac{-1}{\alpha} Ln \frac{13}{22} = \frac{-1}{2.156.10^{-3}} (-0.526) \qquad [4.354]$$

$$t = 244.01 s \qquad [4.355]$$

The Mongol family will suffer from the cold.

5) Thermal science of sport

a) Internal temperature in the stationary regime when 10 cyclists are pedaling

Relationship between flow and temperature:

$$\Phi = 130.9\,(T_i - T_e) \qquad [4.356]$$

$$\Phi = 10x400 = 4,000\ W \qquad [4.357]$$

$$T_i - T_e = \frac{4,000}{130.9} = 30.56°C \qquad [4.358]$$

$$T_i = 30.56° - 8 = 22.56°C \qquad [4.359]$$

b) It is a good thing that extra heating was not installed inside the yurt. But after their efforts, the cyclists should not stay for too long inside the yurt because it will cool down.

5

Non-stationary Conduction

In non-stationary conduction problems, we must take into account changes in temperature over time. The problem can be very complex.

Indeed, it is the temperature that is the solution to the complete heat equation:

$$\rho c \frac{\partial T}{\partial t} = div\left(\lambda \, grad\vec{d} \, T\right) \tag{5.1}$$

the temperature field T is therefore a function of four variables, three in space and one in time, depending, for example, on the geometry of the problem: $T = T(x, y, z, t)$ or $T = T(r, \theta, z, t)$. We say that the problem is a four-dimensional problem.

In addition to this, generally, the physical properties of the material can vary.

We understand that the most complex cases will require a digital approach. In this chapter, we will focus on an analytical approach. To do this, we will consider single-dimensional problems in plane geometry.

5.1. Single-dimensional problem

We consider an environment of semi-infinite (undefined) length that has a plane interface with its exterior. The problem is presumed to be single-dimensional. In particular, on all planes parallel to the interface, the temperature is considered at all times to be constant. We locate the position on an Ox axis. The temperature field at an instant t will therefore be given by an expression $T = T(x,t)$ that must be determined as a function of the conditions imposed at the entry interface.

5.1.1. *Temperature imposed at the interface at instant* $t = 0$

This is the prototype problem in non-stationary conduction. For this reason, we will focus more on it and will provide a full analysis leading to a classic result that is part of the "basic culture" of all thermal scientists.

REMARK.– A less well-informed reader, or one who is allergic to mathematics, should not worry. Simply knowing the expression for the temperature field obtained in the following paragraph, they will understand many examples proposed below.

The environment is presumed homogeneous with constant physical properties as the temperature varies, in particular the volumetric mass ρ, thermal conductibility λ and specific heat capacity c_p.

REMARK.– Again we note that a parameter derived from the previous chapters has appeared, and that this will have a pivotal role in all non-stationary conduction problems, thermal diffusivity a, defined by:

$$a = \frac{\lambda}{\rho c_p} \quad [5.2]$$

a plays a key role and is fundamental in problems with constant physical properties.

In all of these types of problems, we often introduce a *reduced* temperature $\theta(x,t)$ that can be defined in different ways.

We will only require that this *reduced* temperature is a linear function of T, in other words, explicitly in the form

$$\theta(x,t) = \frac{T(x,t) - T_1}{T_2 - T_3} \quad [5.3]$$

where T_1, T_2 and T_3 are temperatures, and which are not all necessarily different, selected depending on the problem in question, and, in particular, on its limit conditions.

We note that the term *reduced* applies, in principle, to adimensional variables. We will also find *derived* variables that will retain their dimension as a temperature.

In this problem, we consider an environment at the initial temperature T_i. We impose at the initial moment in time a temperature T_e at the interface between the environments and the exterior.

Determination of the temperature field $T = T(x,t)$ is a question of solving the single-dimensional heat equation, which is written:

$$\frac{\partial T}{\partial t} = a\,\frac{\partial^2 T}{\partial x^2} \tag{5.4}$$

with which we associate the limit conditions:

$$T(0,t) = T_e \tag{5.5}$$

$$T(x,0) = T_i \tag{5.6}$$

$$x \to \infty \quad ; \quad T \to T_i \tag{5.7}$$

REMARK.– We should recall that the heat equation results from a thermal balance and the limit conditions from the specific conditions of an experiment.

We will introduce an unknown function that takes the form of a reduced temperature, which is linear with unknown temperature, defined by:

$$\theta(x,t) = \frac{T(x,t) - T_e}{T_i - T_e} \tag{5.8}$$

The problem is rewritten as:

$$\frac{\partial \theta}{\partial t} = a \frac{\partial^2 \theta}{\partial x^2} \quad [5.9]$$

with which the following limit conditions are associated:

$$\theta(0,t) = 0 \quad [5.10]$$

$$\theta(x,0) = 1 \quad [5.11]$$

$$x \to \infty \quad ; \quad \theta \to 1 \quad [5.12]$$

We are looking for a solution where the reduced temperature depends exclusively on a composite variable

$$\xi = \frac{x}{2\sqrt{at}} = \frac{x}{\sqrt{4at}} \quad [5.13]$$

Of the two ways of writing it, we generally prefer the second.

In other words,

$$\theta(x,t) = \theta(\xi) \quad [5.14]$$

REMARK.– This variable does not appear intuitively. We can look for its expression in the form $\theta(x,t) = \alpha x^m t^n$. Here, we do not give the calculation and advise curious readers to work through it themselves.

The heat equation, which initially has two variables, will be transformed into an equation with a single variable. We will therefore go from a written expression in the form ∂ to a written expression in d.

The derivatives are transformed as follows:

$$\frac{\partial \theta(x,t)}{\partial t} = \frac{d\theta(\xi)}{d\xi} \frac{\partial \xi}{\partial t} \quad [5.15]$$

$$\frac{\partial\theta(x,t)}{\partial x}=\frac{d\theta(\xi)}{d\xi}\frac{\partial\xi}{\partial x} \qquad [5.16]$$

With:

$$\frac{\partial\xi}{\partial t}=\frac{\partial\xi}{\partial t}\frac{1}{\sqrt{4a}}t^{-\frac{1}{2}}=\frac{1}{\sqrt{4a}}\left(-\frac{1}{2}t^{-\frac{3}{2}}\right)=\frac{-1}{2\sqrt{4at}}\frac{1}{t}=\frac{-\xi}{2t} \qquad [5.17]$$

$$\frac{\partial\xi}{\partial x}=\frac{\partial\xi}{\partial x}\frac{x}{\sqrt{4at}}=\frac{1}{\sqrt{4at}} \qquad [5.18]$$

The terms of the equation are transformed into:

$$\frac{\partial\theta(x,t)}{\partial t}=\frac{-\xi}{2t}\frac{d\theta(\xi)}{d\xi} \qquad [5.19]$$

$$\frac{\partial\theta(x,t)}{\partial x}=\frac{1}{\sqrt{4at}}\frac{d\theta(\xi)}{d\xi} \qquad [5.20]$$

$$\frac{\partial^2\theta(x,t)}{\partial x^2}=\frac{\partial}{\partial x}\left[\frac{1}{\sqrt{4at}}\frac{d\theta(\xi)}{d\xi}\right]=\frac{1}{\sqrt{4at}}\frac{\partial}{\partial x}\left[\frac{d\theta(\xi)}{d\xi}\right]$$

$$=\frac{1}{\sqrt{4at}}\left[\frac{1}{\sqrt{4at}}\frac{d}{d\xi}\frac{d\theta(\xi)}{d\xi}\right] \qquad [5.21]$$

$$\frac{\partial^2\theta(x,t)}{\partial x^2}=\frac{1}{4at}\frac{d^2\theta(\xi)}{d\xi^2} \qquad [5.22]$$

And the equation becomes:

$$\frac{-\xi}{2t}\frac{d\theta(\xi)}{d\xi}=a\frac{1}{4at}\frac{d^2\theta(\xi)}{d\xi^2} \qquad [5.23]$$

We will note, for the purposes of simplifying the expression:

$$\theta' = \frac{d\,\theta(\xi)}{d\,\xi} \qquad [5.24]$$

$$\theta'' = \frac{d^2\theta(\xi)}{d\,\xi^2} \qquad [5.25]$$

which can be written as:

$$-\xi\,\theta' = a\frac{2t}{4at}\,\theta'' \qquad [5.26]$$

or even:

$$\theta'' + 2\xi\,\theta' = 0 \qquad [5.27]$$

The solution to this equation will be obtained by integrating twice, which leads to the appearance of two constants. These constants are determined, as a function of the problem in question, by the limit conditions.

$$\theta'' - 2\xi\,\theta' = 0 \qquad [5.28]$$

integrates into:

$$Ln\,\theta' = -\ \xi^2 + Ln\,C_1 \qquad [5.29]$$

We have chosen to express the constant by a logarithm to simplify the following expression:

$$\theta' = C_1\,\exp-\ \xi^2 \qquad [5.30]$$

$$\theta = C_1\int_O^{\xi}\exp - u^2\,du\ + C_2 \qquad [5.31]$$

REMARK.– For practical reasons, we have chosen to express the solution using a defined integral with a lower limit that is arbitrarily zero. This lower limit will not be zero for all problems and is corrected by the constant.

The expression found for $\theta(\xi)$ encourages us to remind ourselves of the definition of the error function

$$erf\ \xi = \frac{2}{\sqrt{\pi}} \int_0^{\xi} \exp - u^2\, du \qquad [5.32]$$

in addition to the property:

$$\int_0^{\infty} \exp - u^2\, du = \frac{\sqrt{\pi}}{2} \qquad [5.33]$$

which ensures that the error function is normalized to 1.

The limit conditions are written using the variable

$$\xi = \frac{x}{\sqrt{4at}} \qquad [5.34]$$

Application of the limit conditions indicates to us that:

$$\theta(0) = 0 \qquad [5.35]$$

$$\theta(\infty) = 1 \qquad [5.36]$$

Hence:

$$\theta = C_1 \int_O^{\xi} \exp - u^2\, du + C_2 \qquad [5.37]$$

$$0 = C_1 \int_O^{0} \exp - u^2\, du + C_2 \quad \Rightarrow \quad C_2 = 0 \qquad [5.38]$$

$$1 = C_1 \int_O^{\infty} \exp - u^2\, du + C_2 \qquad [5.39]$$

Hence:

$$1 = C_1 = \frac{1}{\int_O^{\infty} \exp - u^2\, du} = \frac{2}{\sqrt{\pi}} \qquad [5.40]$$

The reduced temperature is therefore given by:

$$\theta = \frac{2}{\sqrt{\pi}} \int_{O}^{\xi} \exp - u^2 du \;\; = erf(\xi) \qquad [5.41]$$

We can, if required, re-express $T(x,t)$ as:

$$T(x,t) = (T_i - T_e)\, erf\left(\frac{x}{\sqrt{4at}}\right) + T_e \qquad [5.42]$$

To illustrate this result, we show the evolution of the temperature in Figure 5.1 in a material with thermal diffusivity $a = 5.5.10^{-6}\ m.s^{-2}$

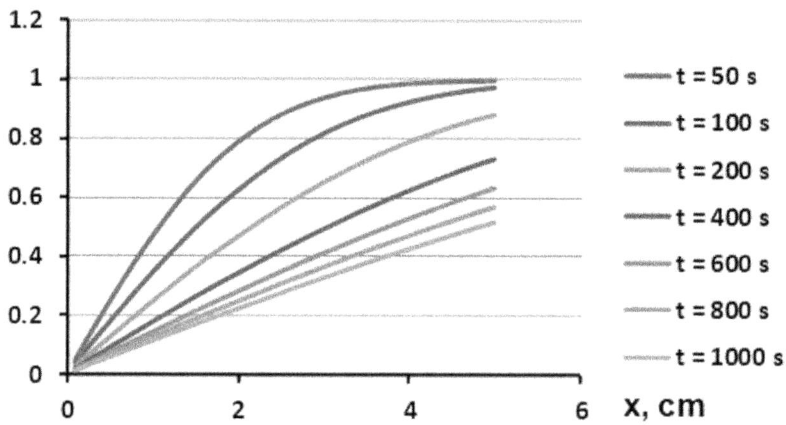

Figure 5.1. *Evolution of a "thermal ladder". Representation of the reduced temperature.* $\theta(x,t) = \frac{T(x,t)-T_1}{T_2-T_3}$ *For a color version of this figure, see www.iste.co.uk/ledoux/heat1.zip*

5.2. Non-stationary conduction with constant flow density

This problem calls into question a wall that has a plane face. The physical properties of the material are presumed invariable, that is, following the usual notation of this book, ρ, c, λ.

We impose a constant thermal flow density φ_0 on the plane face. The initial temperature of the material, homogeneous throughout, is T_i.

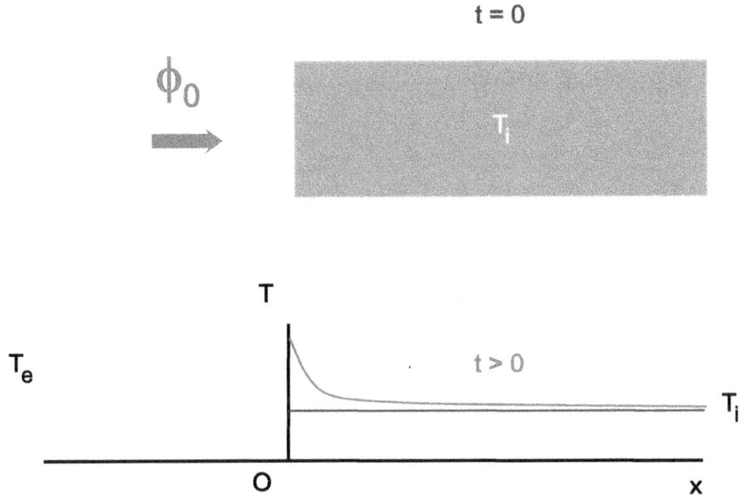

Figure 5.2. *Non-stationary conduction: constant flow density imposed on a face at the initial instant. For a color version of this figure, see www.iste.co.uk/ledoux/heat1.zip*

The transfer is presumed to be single-dimensional (the plane surface is large in comparison with the dimensions of the problem). The temperature will therefore be distributed into isothermal planes, parallel to the contact surface.

We locate these planes on an axis Ox, with origin O on the surface of the wall where the temperature is imposed. The wall is of indefinite length. Its local temperature (on a given plane) is $T = T(x,t)$.

The problem is characterized, as always, by a local balance equation expressed by the heat equation. Here it is in a single-dimensional non-stationary form, associated with the limit conditions that determine the specific conditions.

Hence, the equation:

$$\frac{\partial T}{\partial t} = a\frac{\partial^2 T}{\partial x^2} \qquad [5.43]$$

where a is the thermal diffusivity of the material,

$$a = \frac{\lambda}{\rho c_p} \qquad [5.44]$$

The limit conditions express that the temperature away from the surface of the wall is not affected by the modulation; the second expresses the temperature condition imposed.

In other words:

$$T(\infty, t) = T_i \qquad [5.45]$$

$$-\lambda \left.\frac{\partial T}{\partial x}\right|_{x=0} = \varphi_0 , \qquad [5.46]$$

Here, we give the solution, which is, this time, in dimensional form:

$$T(x,t) - T_i = \frac{2\varphi_0}{\lambda}\sqrt{at}\ ierfc\left(\frac{x}{\sqrt{4at}}\right) \qquad [5.47]$$

Here, we see a function $ierfc\,\xi$ appear, where $\xi = \frac{x}{\sqrt{4at}}$, defined by:

$$ierfc\,\xi = \frac{1}{\sqrt{\pi}}\exp(-\xi^2) - \xi\, erfc\,\xi \qquad [5.48]$$

where $erfc\,\xi$ is the complementary function of $erf\,\xi$, defined by:

$$erfc\,\xi = 1 - erf\,\xi \qquad [5.49]$$

The functions *erf* ξ, *erf* ξ and *ierfc* ξ are tabulated in Appendix 5. These three functions are represented in Figure 5.3 as well as the Appendix:

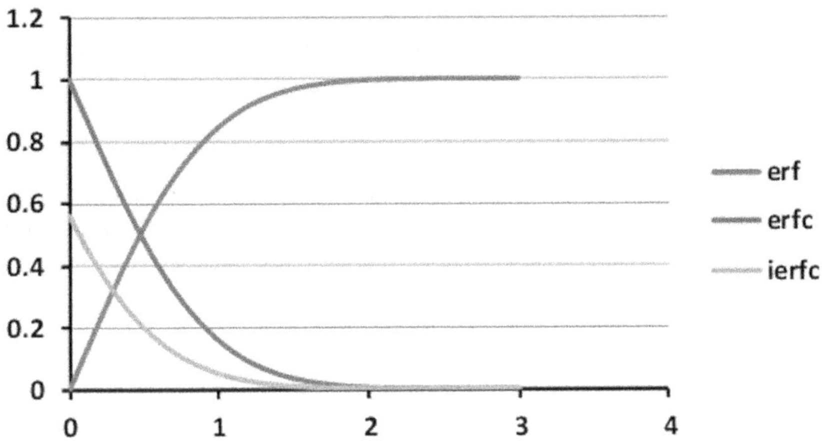

Figure 5.3. *The functions* $erf\,\xi$, $erf\,\xi$ *and* $ierfc\,\xi$

5.3. Temperature imposed on the wall: sinusoidal variation

This problem involves a wall with a plane face. The physical properties of the material are presumed invariable, that is, following the usual notations of this book, ρ, c, λ.

Here, we do not consider a problem beginning at an initial instant in time, but instead the results of a temperature variable on the transfer between the outside and the material.

This temperature will have a fixed value to which a sinusoidal variation is added.

In this type of problem, when we impose this type of condition, we have first a transitory solution, then an established regime. Here, we are interested in this established regime.

The transfer is presumed to be single-dimensional (the plane surface is large with regard to the dimensions of the problem). The temperature will therefore be distributed in isothermal planes, parallel to the contact surface.

We locate these planes on an axis Ox, with origin O on the surface of the wall where the temperature is imposed. The wall is of an undefined length. Its local temperature (on a given plane) is $T = T(x,t)$.

The problem is characterized, as always, by a local balance equation expressed by the heat equation. Here it is in the single-dimensional non-stationary form, associated with the limit conditions that determine the specific conditions.

In other words, the equation:

$$\frac{\partial T}{\partial t} = a \frac{\partial T}{\partial x^2} \qquad [5.50]$$

where a is the thermal diffusivity of the material,

$$a = \frac{\lambda}{\rho c_p} \qquad [5.51]$$

The limit conditions express that the temperature away from the surface of the wall is not affected by modulation; the second expresses the imposed temperature condition.

In other words:

$$T(\infty, t) = T_i \qquad [5.52]$$

$$T(0,t) = T_i + T_0 \cos \omega t, \qquad [5.53]$$

where T_0 is a modulation amplitude and ω is a pulsatance.

Figure 5.4. *Non-stationary conduction: sinusoidal temperature on a face*

A new relative temperature function is defined:

$$\theta = T - T_i \tag{5.54}$$

The problem becomes

$$\frac{\partial \theta}{\partial t} = a \frac{\partial^2 \theta}{\partial x^2} \tag{5.55}$$

associated with

$$\theta(\infty, t) = 0 \tag{5.56}$$

$$\theta(0, t) = T_0 \cos \omega t \tag{5.57}$$

We will use a complex formulation to solve this: we look for the reduced temperature in a complex form; the solution will be the real value of this complex function.

REMARK.– This process is currently used in classes on electricity. Readers who are not familiar with this method should refer to these.

Therefore, we write:

$$\theta(0, t) = T_0 \, e^{i \omega t} \tag{5.58}$$

And we look for $\theta(x,t)$ in the form:

$$\theta(x, t) = f(x) \, e^{i \omega t} \tag{5.59}$$

This form, separating terms in x from terms in t, expresses that we are in an "established regime".

Along the same lines:

$$\theta(\infty,t)=0 \quad [5.60]$$

$$\theta(0,t)=T_0\, e^{i\omega t} \quad [5.61]$$

The solution will be the real part of the result that is found.

The terms of the equation can easily be deduced:

$$\frac{\partial \theta}{\partial t}= i\,\omega f\, e^{i\omega t} \quad [5.62]$$

$$\frac{\partial^2 \theta}{\partial x^2} = f''e^{i\omega t} \quad [5.63]$$

Hence the equation:

$$i\,\omega f\, e^{i\omega t} = a\, f''e^{i\omega t} \quad [5.64]$$

$$\left(f''+\frac{i\,\omega f}{a}\right)e^{i\omega t}=0 \quad [5.65]$$

Leading to:

$$f''+\frac{i\,\omega f}{a}=0 \quad [5.66]$$

Which has the solution:

$$f(x)=A\exp\sqrt{\frac{i\,\omega}{a}}\,x+B\exp-\sqrt{\frac{i\,\omega}{a}}\,x \quad [5.67]$$

where A and B are two constants.

Here, we see the square root of i appear, an imaginary number that is already equal to $i=\sqrt{-1}$.

REMARK.– $\sqrt{i}$ is not necessarily familiar to all readers. We will easily find its value by writing it in the form of a complex number where a and b are real numbers to be found:

$$\sqrt{i}=a+ib \qquad [5.68]$$

$$i=\left(\sqrt{i}\right)^2=\left(a+ib\right)^2=a^2+b^2+2iab \qquad [5.69]$$

by identification:

$$0+i=a^2-b^2+2iab \qquad [5.70]$$

$$a=b=\frac{1}{\sqrt{2}}$$

Therefore:

$$\sqrt{i}=\frac{1+i}{2} \qquad [5.71]$$

The limit conditions allow the following to be written:

at $x=0$: [5.72]

$$f=0 \qquad \Rightarrow \qquad A+B=T_0 \qquad [5.73]$$

at infinity:

$$x\rightarrow\infty \qquad \Rightarrow \qquad f\rightarrow 0 \qquad [5.74]$$

The function therefore becomes:

$$f(x)=A\exp\sqrt{\frac{i\omega}{a}}x+B\exp-\sqrt{\frac{i\omega}{a}}x=A\exp\frac{1+i}{\sqrt{2}}\sqrt{\frac{\omega}{a}}x+B\exp-\frac{1+i}{\sqrt{2}}\sqrt{\frac{\omega}{a}}x$$

$$= A\left(\exp\sqrt{\frac{\omega}{2a}}x\right)\left(\exp i\sqrt{\frac{\omega}{2a}}x\right) + B\left(\exp -\sqrt{\frac{\omega}{2a}}x\right)\left(\exp -i\sqrt{\frac{\omega}{2a}}x\right) \qquad [5.75]$$

which can only be zero at infinity for: $A = 0$.

Therefore:

$$B = T_0 \qquad [5.76]$$

Finally:

$$\theta(x,t) = f(x)\, e^{i\omega t} = T_0 \exp -\sqrt{\frac{i\omega}{a}}x \exp i\omega t \qquad [5.77]$$

Again explaining $\sqrt{i}$:

$$\theta(x,t) = T_0 \exp -\frac{1+i}{\sqrt{2}}\sqrt{\frac{\omega}{a}}x \exp i\omega t = T_0 \exp -\sqrt{\frac{\omega}{2a}}x \exp i\left(\omega t - \sqrt{\frac{\omega}{2a}}x\right) \qquad [5.78]$$

$$\theta(x,t) = T_0 \exp -\sqrt{\frac{\omega}{2a}}x \exp i\left(\omega t - \sqrt{\frac{\omega}{2a}}x\right)$$

$$= T_0 \exp -\sqrt{\frac{\omega}{2a}}x\left[\cos\left(\omega t - \sqrt{\frac{\omega}{2a}}x\right) + i\sin\left(\omega t - \sqrt{\frac{\omega}{2a}}x\right)\right] \qquad [5.79]$$

Taking the real part, the solution to our problem:

$$\theta(x,t) = T_0 \exp\left(-\sqrt{\frac{\omega}{2a}}x\right)\cos\left(\omega t - \sqrt{\frac{\omega}{2a}}x\right) \qquad [5.80]$$

The final expression of $T(x,t)$ comes out as:

$$T(x,t) = T_i + T_0 \exp\left(-\sqrt{\frac{\omega}{2a}}x\right)\cos\left(\omega t - \sqrt{\frac{\omega}{2a}}x\right) \quad [5.81]$$

We see that the temperature at a point of the material is the sum of the initial constant temperature and of a modulation.

This modulation has an amplitude and a phase that depends on the considered plane:

on the x axis plane, the amplitude is:

$$T_0 \exp\left(-\sqrt{\frac{\omega}{2a}}x\right) \quad [5.82]$$

The phase is in this same plane:

$$\Phi = \sqrt{\frac{\omega}{2a}}x \quad [5.83]$$

This problem has various practical applications; it allows us to understand how the variation of the external temperature can act on the temperature of a wall, in line with alternating day–night temperatures for a thin wall, or seasonal variations for a thick wall (in the case of the old walls of farms or thatched cottages, barns and so on etc.).

Another application is in the metrology of thermal materials, in particular, insulating materials, for which the method presented at the beginning of the chapter may prove to be tricky. We apply a modulated temperature on the surface of a material and we place thin probes in different planes at a distance from this surface. By measuring the temperature in different locations, we can measure the thermal diffusivity, using the amplitude, or even better, using the phase shift.

5.4. Problem with two walls stuck together

This problem involves two materials that have a plane face.

These two materials are denoted by 1 and 2 ; they are initially at the respective initial temperatures T_{1i} and T_{2i}. Their respective physical properties, which are supposedly invariable, are ρ_1, c_{P1}, λ_1 and ρ_2, c_{P2}, λ_2 when written using the usual notations of this book.

At the instant $t=0$, these two materials are stuck together by a plane face. The transfer of non-stationary heat is studied through this contact surface.

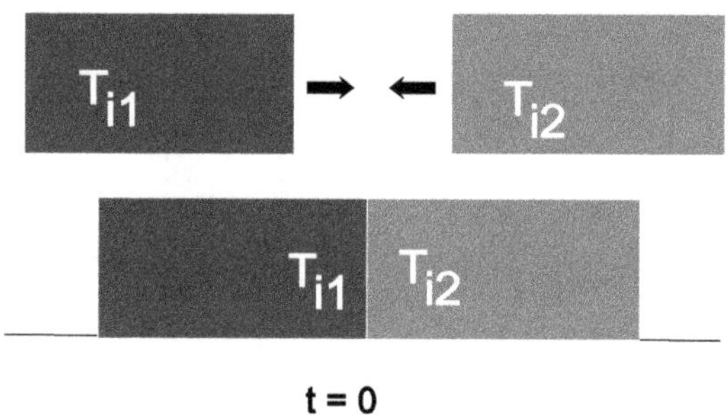

Figure 5.5. *Non-stationary conduction: contact of two bodies at different temperatures. For a color version of this figure, see www.iste.co.uk/ledoux/heat1.zip*

The transfer is presumed to be single-dimensional (the plane surface is large in comparison to the dimensions of the problem). The temperature will therefore be distributed in isothermal planes, parallel to the contact surface.

We locate these planes on an axis Ox, with origin O on the separation surface. Material 1 will correspond to the negative axes (conventionally to the left of the origin) and material 2 to the positive axes (conventionally to the right of the origin).

To structure this in an equation, we will presume that these environments extend infinitely to the left and right. The respective temperatures of these two materials will be indicated by:

$$T_1 = T_1(x,t) \text{ and } T_2 = T_2(x,t). \qquad [5.84]$$

Therefore, contrary to the other problems seen here, the solution lies in two temperature functions.

The problem is characterized, as always, by a local balance equation expressed by the heat equation. Here it is in the non-stationary single-dimensional form, associated with limit conditions that determine the specific conditions.

These two systems need to be written down for each environment, where each differential equation is only valid for half of the space.

So we have the system of equations:

In environment 1,

$$x < 0 \qquad \frac{\partial T_1}{\partial t} = a_1 \frac{\partial T_1}{\partial x} \qquad [5.85]$$

In environment 2,

$$x > 0 \qquad \frac{\partial T_2}{\partial t} = a_2 \frac{\partial T_2}{\partial x} \qquad [5.86]$$

where a is the thermal diffusivity of the materials,

$$a = \frac{\lambda}{\rho c_p} \qquad [5.87]$$

We will define two temperature variables, which are not reduced here:

$$\theta_1 = T_1 - T_{1i} \text{ and } \theta_2 = T_2 - T_{2i} \qquad [5.88]$$

The system to be resolved becomes:

In environment 1,

$$x < 0 \qquad \frac{\partial \theta_1}{\partial t} = a_1 \frac{\partial \theta_1}{\partial x} \qquad [5.89]$$

In environment 2,

$$x > 0 \qquad \frac{\partial \theta_2}{\partial t} = a_2 \frac{\partial \theta_2}{\partial x} \tag{5.90}$$

where a is the thermal diffusivity of the materials,

$$a = \frac{\lambda}{\rho c_P} \tag{5.91}$$

associated with the limit conditions:

In environment 1,

$$x < 0$$

$$\theta_1(x,0) = 0 \tag{5.92}$$

$$\theta_1(-\infty,t) = 0$$

In environment 2,

$$x > 0$$

$$\theta_2(x,0) = 0 \tag{5.93}$$

$$\theta_1(+\infty,t) = 0 \tag{5.94}$$

In addition, at the interface $(x = 0))$, we have a continuity of flows and identity of the temperatures in the two environments:

$$x = 0 \qquad \lambda_1 \frac{\partial T_1}{\partial x} = \lambda_2 \frac{\partial T_2}{\partial x} \tag{5.95}$$

$$x = 0 \qquad T_1(0,t) = T_2(0,t) \tag{5.96}$$

hence, further to this:

$$x = 0 \qquad \lambda_1 \frac{\partial \theta_1}{\partial x} = \lambda_2 \frac{\partial \theta_2}{\partial x} \tag{5.97}$$

$$x = 0 \qquad \theta_1 - \theta_2 = T_{2i} - T_{1i} \tag{5.98}$$

Here, we will not give the details of this quite complex solution, the perspective being to provide an applicable relationship.

The temperatures in the two environments will be given in a non-dimensional form. Logically, we find two elements that appear in the previous problems:

the variable

$$\xi = \frac{x}{2\sqrt{at}} = \frac{x}{\sqrt{4at}} \tag{5.99}$$

the error function, in its complementary form

$$erfc\,\xi = 1 - erf\,\xi \tag{5.100}$$

where

$$erf\,\xi = \frac{2}{\sqrt{\pi}} \int_0^{\xi} \exp - u^2\,du \tag{5.101}$$

In environment 1 "on the left":

$$\frac{T_1(x,t) - T_{i1}}{T_{i2} - T_{i1}} = \frac{E_2}{E_1 + E_2}\, erfc \frac{|x|}{\sqrt{4at}} \tag{5.102}$$

Here, we note the use of an absolute value. Indeed, in this environment, all axes are negative.

In environment 2 “on the right”, we have:

$$\frac{T_2(x,t)-T_{i2}}{T_{i1}-T_{i2}}=\frac{E_1}{E_1+E_2}erfc\frac{x}{\sqrt{4at}} \qquad [5.103]$$

At this point, note that the denominator of the reduced temperature is different in the two cases (influence on the sign).

In these two equations, we see a new parameter E appear, characteristic to each material. It is called **effusivity** and is defined by:

$$E=\lambda\rho c \qquad [5.104]$$

In other words:

$$E_1=\lambda_1\rho_1 c_1 \qquad [5.105a]$$

$$E_2=\lambda_2\rho_2 c_2 \qquad [5.105b]$$

5.5. Application examples

5.5.1. *Simple applications*

EXAMPLE 5.1.– Thermal response of a concrete block

A concrete block with dimensions $60*30*20\ \text{cm}^3$ rests at the initial moment in time on its largest face on a metal plate that is maintained at a constant temperature of 60°C. The initial temperature of the concrete block is 20°C.

The heat transfer is presumed to be single-dimensional, therefore perpendicular to the metal plate.

How long will it take for the center of the concrete block to reach the temperature of 40°C?

The volumetric mass of the concrete is $\rho = 2300\, kg.m^{-3}$, its specific heat capacity is $c = 878\, J.kg^{-1}.K^{-1}$ and its thermal conductibility is $\lambda_B = 1.4 W.m^{-1}.K^{-1}$.

SOLUTION.–

We are faced with a classic problem of non-stationary single-dimensional transfer, with a temperature of $T_e = 60\,°C$ imposed on a face of a solid, initially at a homogeneous temperature

$$T_i = 20\,°C\,. \qquad [5.106]$$

We make the change unknown:

$$\theta = \frac{T(x,t) - T_e}{T_i - T_e} \qquad [5.107]$$

The solution is classic.

We define the composite variable:

$$\xi = \xi(x,t) = \frac{x}{2\sqrt{at}} = \frac{x}{\sqrt{4at}} \qquad [5.108]$$

we then know that:

$$\theta = erf(\xi) \qquad [5.109]$$

where

$$erf(\xi) = \frac{2}{\sqrt{\pi}} \int_0^{\xi} e^{-u^2} du \qquad [5.110]$$

is the error function, for which a table is provided in Appendix 5.

Let us calculate the thermal diffusivity of the block:

$$a = \frac{\lambda}{\rho c} = \frac{1.4}{2300 * 878} = 6.93.10^{-7}\ m^2.s^{-1} \quad [5.111]$$

If the temperature is $T = 40\,°C$ at this point, we have a value of

$$\theta = \frac{40 - 60}{20 - 60} = \frac{-20}{-40} = 0.5 \quad [5.112]$$

This value is obtained for

$$\xi = 0.477 \quad [5.113]$$

This value must be found in the middle of the block, in other words, at $x = 0.1\,m$ from the heating face.

We easily deduce the corresponding time:

$$\xi(x,t) = \frac{x}{2\sqrt{at}} = \frac{x}{\sqrt{4at}} \quad [5.114]$$

$$t = \frac{1}{4a}\left(\frac{x}{\xi}\right)^2 \quad [5.115]$$

$$t = \frac{1}{4 * 6.93.10^{-7}}\left(\frac{0.1}{0.477}\right)^2 \quad [5.116]$$

$$t = 15855\,s = 4\,h\,24\,mn\,15\,s \quad [5.117]$$

EXAMPLE 5.2.– Thermal response of a steel cube

We plunge a large cube of stainless steel into a bath of molten salts. The initial temperature of the metal is 25°C. The temperature of the salt bath is 500°C. This temperature is imposed on the cube faces. Given the size of the cube, we estimate that the transfer of non-stationary heat for each face is single-dimensional (the flow of heat is locally perpendicular to each face).

How long will it take for the temperature to reach 400°C at 2 cm from the surface of each face of the cube?

Data for stainless steel: $a = 5.53.10^{-6}\ m.s^{-2}$.

SOLUTION.–

We are facing a classic problem of single-dimensional non-stationary transfer, with a temperature $T_e = 60\,°C$ imposed on a face of a solid, initially at a homogeneous temperature $T_i = 20\,°C$.

We implement the change unknown:

$$\theta = \frac{T(x,t) - T_e}{T_i - T_e} \tag{5.118}$$

The solution is classic.

We define the composite variable:

$$\xi = \xi(x,t) = \frac{x}{2\sqrt{at}} = \frac{x}{\sqrt{4at}} \tag{5.119}$$

we then know that:

$$\theta = erf(\xi) \tag{5.120}$$

where $erf(\xi) = \frac{2}{\sqrt{\pi}} \int_0^{\xi} e^{-u^2} du$ is the error function, for which a table is provided in Appendix 5.

If the temperature is $T = 400\,°C$ at this point, we have a value of

$$\theta = \frac{400 - 500}{25 - 500} = 0.21 \tag{5.121}$$

This value is obtained for

$$\xi = 0.188 \qquad [5.122]$$

This value must be found at $x = 2.10^{-2}\ m$ from the wall in contact with the molten salt.

We can easily deduce the corresponding time:

$$\xi(x,t) = \frac{x}{2\sqrt{at}} = \frac{x}{\sqrt{4at}} \qquad [5.123]$$

$$t = \frac{1}{4a}\left(\frac{x}{\xi}\right)^2 \qquad [5.124]$$

$$t = \frac{1}{4 * 5.53.10^{-6}}\left(\frac{0.021}{0.188}\right)^2 \qquad [5.125]$$

$$t = 512\,s = 8\,mn\,3\,s \qquad [5.126]$$

EXAMPLE 5.3.– Thermal response to a concrete wall

We consider the external layer of a wall. This layer is made of agglomerated concrete with a thickness of 20 cm. The thermal diffusivity of this concrete is $a = 8.38.10^{-7}\ m.s^{-2}$.

The wall is initially at a homogeneous temperature of 15°C.

The **external** temperature is suddenly brought down to 0°C.

How long will it take for the temperature to fall to 5°C at a distance of 5 cm from the **internal** face of the concrete wall? Express this time in hours and minutes.

SOLUTION.–

We are facing a classic problem of single-dimensional non-stationary transfer, with a temperature T_e imposed on a face of a solid, initially at a homogeneous temperature T_i.

The change unknown is made:

$$\theta = \frac{T(x,t) - T_e}{T_i - T_e} \quad [5.127]$$

The solution is classic.

We define the composite variable:

$$\xi = \xi(x,t) = \frac{x}{2\sqrt{at}} = \frac{x}{\sqrt{4at}} \quad [5.128]$$

we then know that:

$$\theta = erf(\xi) \quad [5.129]$$

where $erf(\xi) = \frac{2}{\sqrt{\pi}} \int_0^{\xi} e^{-u^2} du$ is the error function, for which a table is provided in Appendix 5.

If the temperature is $T = 5°C$, we have, at this point, a value of

$$\theta = \frac{5 - 0}{15 - 0} = 0.333 \quad [5.130]$$

This value is obtained for $\xi = 0.305$.

This value must be found at $x = 20 - 5 = 15\,cm = 15.10^{-2}\,m$ from the face of the wall in contact with the molten salt.

We easily deduce the corresponding time from this:

$$\xi(x,t)=\frac{x}{2\sqrt{at}}=\frac{x}{\sqrt{4at}} \quad [5.131]$$

$$t=\frac{1}{4a}\left(\frac{x}{\xi}\right)^2 \quad [5.132]$$

$$t=\frac{1}{4*8.38.10^{-7}}\left(\frac{0.15}{0.305}\right)^2 \quad [5.133]$$

$$t=72157\ s=20\,h\ 2\,mn\ 37\,s \quad [5.134]$$

EXAMPLE 5.4.– Thermal response of an immersed cube

We dip a cube of stainless steel ($18\%Cr\,/\,8\%Ni$) with sides of 14 cm, initially at an atmospheric pressure of 21°C, into a container filled with boiling water at 95°C.

We will suppose (approximation) that the non-stationary thermal transfer observed is single-dimensional with respect to each of the faces.

We will also presume that over the course of the entire transfer, the water will impose a temperature of 95°C onto each face of the cube.

We provide the thermal properties of the steel grade used:

Density:

$$\rho=7816\ kg.m^{-3}$$

Thermal conductibility:

$$\lambda=16.3\ kg.m^{-3}\ W\ m^{-1}\ K^{-1}$$

Specific heat capacity:

$$c=460\ J.kg^{-1}$$

1) Evaluate in the context of this approximation how long it will take for the temperature to reach 55°C at $x = 4\,cm$ from each of the faces.

2) Evaluate by the same process (which then becomes quite large) how long it will take for the temperature to reach 80°C at the center of the cube.

SOLUTION.– **1) Here, we see a problem of variable conduction at an imposed wall temperature**. We therefore know that the reduced temperature will be of the form:

$$\theta = erf\left(\xi\right) \qquad [5.135]$$

with

$$\theta = \frac{T\left(x,t\right) - T_e}{T_i - T_e} \qquad [5.136]$$

where T_e is the imposed temperature and T_i is the initial temperature of the cube.

Prior to all problems of this type, we will calculate the thermal diffusivity of the steel

$$a = \frac{\lambda}{\rho c} = \frac{16.3}{7816 * 460} = 4.53.10^{-6}\ m^2.s^{-1} \qquad [5.137]$$

we fix

$$T = 55\ °C \ \text{ at } \ x = 4.10^{-2}\ m \qquad [5.138]$$

the time, which is the value that we are looking for, will be deduced from the value of giving θ as prescribed.

In other words:

$$\theta = \frac{55 - 95}{21 - 95} = 0.54 \qquad [5.139]$$

Consulting the table of $erf(\xi)$, by interpolation, we find:

$$0.54 = erf(0.5633) \qquad [5.140]$$

From this, we deduce the time that we are looking for

$$\xi(x,t) = \frac{x}{2\sqrt{at}} = 0.523 = \frac{4.10^{-2}}{\sqrt{4 * 4.53.10^{-6} * t}} \qquad [5.141]$$

$$t = \left(\frac{4.10^{-2}}{0.523}\right)^2 \frac{1}{4 * 4.53.10^{-6}} \qquad [5.142]$$

$$t = 322.8\ s = 5\ mn\ 23\ s \qquad [5.143]$$

2) The approach followed in the previous question will be used again here

At the center of the cube

$$x = 7\ cm = 7.10^{-2}\ m \qquad [5.144]$$

For 80°C at the center of the cube, θ has the value:

$$\theta = \frac{80-95}{21-95} = 0.2027 \qquad [5.145]$$

So, in the same way as before:

$$0.2027 = erf(0.182) \qquad [5.146]$$

So, the time:

$$t = \left(\frac{7.10^{-2}}{0.182}\right)^2 \frac{1}{4 * 4.53.10^{-6}} \qquad [5.147]$$

$$t = 8164\ s = 2\ hr\ 16\ mn\ 4\ s \qquad [5.148]$$

5.5.2. *Some scenes from daily life*

EXAMPLE 5.5.– Problem of thermal phenomena on a daily basis: ironing day

A housewife is ironing her laundry. She inadvertently places her iron on a pile of laundry. The iron has a power of $P = 1000W$, and the surface of the plate (surface in contact with the laundry) is $S = 100cm^2$. The atmospheric temperature at which the laundry is initially found is $T_i = 20°C$.

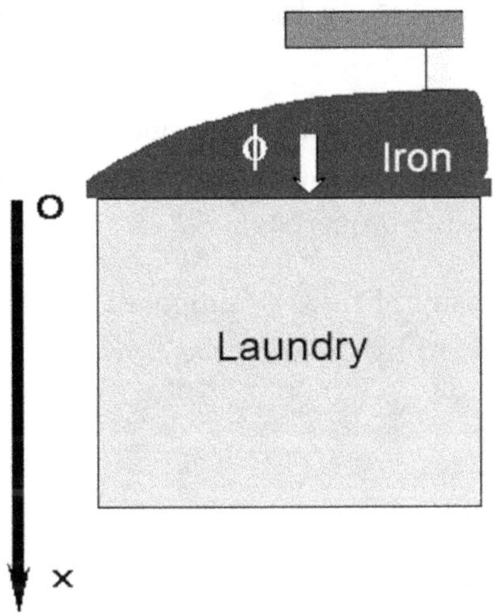

Figure 5.6. *The iron and its pile of laundry*

The pile of laundry will be assimilated to a homogeneous, semi-infinite environment whose physical characteristics are:

Thermal conductibility of the laundry:

$$\lambda = 0.46 W.m^{-1}.kg^{-1}$$

Density:

$$\rho = 929.kg.m^{-3}$$

Specific heat capacity of the laundry:

$$c = 1830 J.kg^{-1}$$

We consider that the polyester material starts to melt at temperatures of $T_F = 240\,°C$:

1) What is the flow density φ_0 imposed by the iron at the top of the pile of laundry?

2) Give the expression as a function of the time taken by the top of the pile of laundry to heat up: $T(x=0,\, t) - T_i$.

How long will it take for the laundry at the top of the pile to begin to melt?

3) What is the temperature at $x = 5\,cm$ under the surface of the pile 30 s after the iron has been placed on the pile? (We ignore the effects of the surface fusion on the textile properties).

SOLUTION.– **1)** The density of flow φ_0 imposed by the iron on the top of the pile of laundry is obtained by dividing the flow $P = 1000\,W$ by the heating surface $S = 100\,cm^2 = 10^{-2}\ m^2$:

$$\varphi_0 = \frac{1000}{10^{-2}} = 10^5\ W\ m^{-2} \qquad [5.149]$$

2) Expression as a function of the time taken for the top of the pile of laundry to heat up:

We are facing a non-stationary problem with imposed constant flow. The temperature is then expressed by:

$$\theta = \frac{2\,\varphi_0}{\lambda}\sqrt{a\,t}\ ierfc\left(\xi\right) \qquad [5.150]$$

with the usual unknown and variable:

$$\theta\left(x,t\right) = T\left(x,t\right) - T_i \qquad [5.151]$$

$$\xi = \frac{x}{2\sqrt{a\,t}} = \frac{x}{\sqrt{4\,a\,t}} \qquad [5.152]$$

We are looking for the value of

$$T(x=0,\ t)-T_i=\theta(0,t) \quad [5.153]$$

In this case,

$$\xi=\frac{0}{2\sqrt{at}}=0 \quad [5.154]$$

$$\theta(0,t)=\frac{2\varphi_0}{\lambda}\sqrt{at}\ ierfc(0) \quad [5.155]$$

We know that

$$ierfc(0)=0.5642 \quad [5.156]$$

We calculate

$$a=\frac{\lambda}{\rho c}=2.7.10^{-7} \quad [5.157]$$

We will reach the temperature $T(0,t)=220\ °C$ at the top of the pile for the time t_F such that:

$$\theta(0,t)=\frac{2\varphi_0}{\lambda}\sqrt{at}\ ierfc(0) \quad [5.158]$$

$$\theta(0,t)=240-20=220=\frac{2.10^5}{0.46}\sqrt{2.7.10^{-7}*t_F}\ *0.5642 \quad [5.159]$$

$$t_F^{0,5}=220=1.72 \quad [5.160]$$

$$t_F=2.9\ s \quad [5.161]$$

It is highly advisable to be very careful!

3) The temperature at $x = 5\,cm$ below the surface of the pile 30 s after the iron is placed on the pile will be calculated using the expression θ for $x = 5.10^{-2}\ m$ and $t = 30\ s$.

Then:

$$\xi = \frac{5.10^{-2}}{2\sqrt{2.7.10^{-7} * 30}} = 0.878 \quad [5.162]$$

$$ierfc(0.878) = 0.073 \quad [5.163]$$

$$\theta = \frac{2\varphi_0}{\lambda}\sqrt{at}\ ierfc(\xi) \quad [5.164a]$$

$$\theta(5.10^{-2}, 30) = \frac{2.10^5}{0.46}\sqrt{2.7.10^{-7} * 30} * 0.073 \quad [5.164b]$$

$$\theta(5.10^{-2}, 30) = 90.33 \quad [5.165]$$

$$T(5.10^{-2}, 30) = 110.33 \quad [5.166]$$

Some of the laundry can be saved…

EXAMPLE 5.6.– The physicist is going to cook an egg

An egg, when it is laid, is made up of a shell, egg white and yolk. The **white of the egg** is a part of the egg, made essentially of an albumin, ovalbumin, which protects the yolk or the zygote.

The albumin presents advantages due to its properties as a coagulant and surfactant (this allows beaten egg whites to hold their shape).

It is **coagulation properties** that we are interested in here.

This coagulation occurs at 57°C.

In the egg yolk, the proteins also coagulate, but at a slightly higher temperature of 65°C.

In the simplified model that follows, it will be considered that an egg is made up of

1) A central cylindrical zone of radius 0.5 cm, known as the egg yolk.

2) An annular cylindrical zone surrounding the first, of thickness 5 mm, called the egg white.

3) An extremely thin shell that will have no effect on the thermal phenomena.

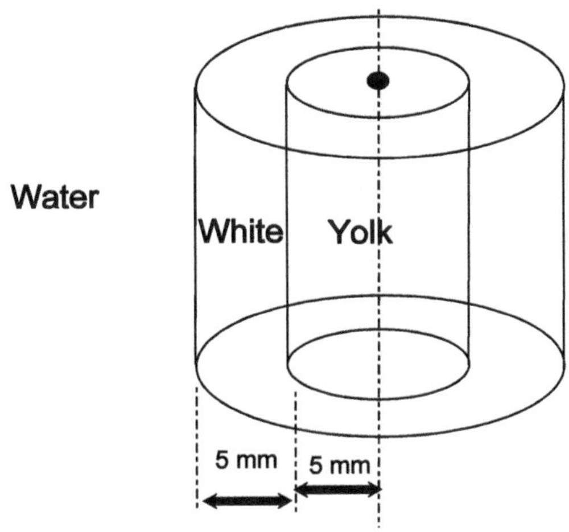

Figure 5.7. *Diagram of an egg*

We will consider that **the white and the yolk of the egg have the same thermal properties**, which are very close to those of albumin. We will therefore consider that the egg white and the yolk constitute a single environment, which is also strictly immobile.

1) Initial calculation

A very thin plane wall, for which we will consider thermal resistance to be zero, separates two spaces containing:

– On the left, water at 100°C, which will only have the effect of maintaining this temperature at the separation for the duration of the problem.

– On the right, albumin, composed mainly of water containing macromolecules. This environment will be considered perfectly immobile, thus excluding all appearance of convection coefficients at the interface.

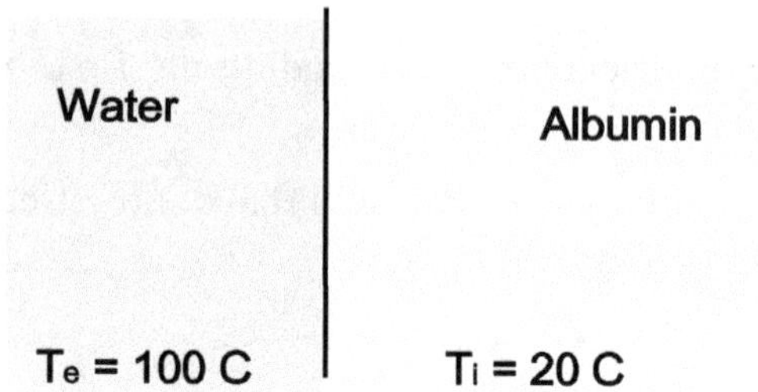

Figure 5.8. *Initial calculation*

The thermal characteristics of the albumin are evaluated to be:

Thermal conductivity:

$$\lambda = 0.6\,Wm^{-2}K^{-2}$$

Density:

$$\rho = 1000\,kg.m^{-3}$$

Specific heat capacity:

$$c = 4180\,J\,kg^{-1}\,K^{-1}.$$

REMARK.– Well-informed readers will certainly not miss the fact that these are the thermal properties of water. It is not always easy to find precise data in terms of the physics and chemistry of food items. Egg white contains 88% water. Among the other constituents of albumin, we find 10.6% of globular proteins, the main one being known as ovalbumin. For this simple model, we have chosen data related to water, which are easier to access.

At an initial instant in time, the temperature of the albumin, in the right-hand compartment, is uniformly equal to $T_i = 20°C$.

a) How long will it take for the temperature to reach $T_1 = 57\,°C$ in the albumin at a distance $e = 0.5\,\text{cm}$ from the separation surface with water at 100°C.

b) How long will it take for the temperature to be equal to $T_2 = 65\,°C$ in the albumin at a distance $e = 1\,cm$ from the separation surface with water at 100°C.

2) Cooking

This question, requiring no additional calculation, only requires a little physical reasoning, which is a highly desirable quality of all thermal scientists.

We will admit that if the part at the center of the egg white has reached a temperature of 57°C, the white will be coagulated and the yolk will remain soft. We will have a soft-boiled egg.

In the same way, if we reach the minimum temperature of 65°C at the center of the yolk, the white and the yolk will coagulate and we will have cooked a hard-boiled egg.

a) Soft-boiled egg

Remembering that both a "soft-boiled" and "hard-boiled" egg are cooked by plunging the egg into boiling water (here presumed to be 100°C), it is only the cooking time that makes the difference.

Modeling the egg geometrically, as we did at the beginning of this text, we will examine how much time is required, using our model, to cook a soft-boiled egg.

REMARK.– The highly non-stationary aspect of the operation explains why it is recommended to take the eggs out of the refrigerator for a sufficient amount of time before cooking, in order to preserve the shell. Here, we have a material problem (thermal shock) that falls outside the remit of our considerations.

b) Hard-boiled egg

In the same way, estimate the order of magnitude of the time required to cook a hard-boiled egg.

c) Are these results coherent with your personal experience (if any)?

If not, in your opinion, why?

SOLUTION.–

1) Initial calculation

We are facing a single-dimensional non-stationary problem with an imposed temperature on a marker in space.

We will determine the thermal diffusivity of the albumin

$$a = \frac{\lambda}{\rho c} = 1.43.10^{-7}\ m^2 s^{-1} \qquad [5.167a]$$

a) We are looking for the time t required so that equal T_1 reaches 57°C in the albumin at a distance $e = 0.5\,cm$ from the separation surface with water at 100°C.

This is a problem with an imposed wall temperature.

We will firstly calculate the thermal diffusivity of the albumin

$$a = \frac{\lambda}{\rho c} = 1.43.10^{-7}\ m^2 s^{-1} \qquad [5.167b]$$

We define a reduced variable:

$$\theta = \frac{T - T_e}{T_i - T_e} \qquad [5.168]$$

with

$$T_i = 20°C \qquad [5.169]$$

$$T_e = 100°C \qquad [5.170]$$

We know that θ will be of the form

$$\theta = erf\ \xi \qquad [5.171]$$

with:

$$\xi = \frac{x}{\sqrt{4\,a\,t}} \qquad [5.172]$$

We write

$$T\left(x,t\right) = 57°C \qquad [5.173]$$

then:

$$\theta = 0.537 \qquad [5.174]$$

and the table gives us:

$$\xi = 0.519 \qquad [5.175]$$

the following is then deduced:

$$\xi = \frac{5.10^{-3}}{\sqrt{4 * 1.43.10^{-7}\,t}} = \frac{6.61}{\sqrt{t}} \qquad [5.176]$$

$$\sqrt{t} = 12.7 \qquad [5.177]$$

$$t = 162\,s = 2.7\,mn \qquad [5.178]$$

b) We now look for the moment when the temperature T_2 reaches 65°C in the albumin at a distance of $e = 1\,cm$ from the separation surface with the water at 100°C.

The calculation will be based on the previous one:

$$\theta = \frac{T - T_e}{T_i - T_e} \qquad [5.179]$$

$$T_i = 20°C \quad [5.180]$$

$$T_e = 100°C \quad [5.181]$$

$$\theta = erf\ \xi \quad [5.182]$$

$$\xi = \frac{x}{\sqrt{4at}} \quad [5.183]$$

$$T(x,t) = 57°C \quad [5.184]$$

$$\theta = 0.437 \ ; \ \xi = 0.413$$

$$\xi = \frac{10^{-2}}{\sqrt{4 * 1.43.10^{-7}\, t}} = \frac{13.22}{\sqrt{t}} \quad [5.185]$$

$$\sqrt{t} = 32.01 \quad [5.186]$$

$$t = 1026\, s = 17\, mn \quad [5.187]$$

2) Cooking an egg

a) Soft-boiled egg

The preliminary calculations give us the result: using the modeling in the albumin, looking for a temperature of 57°C at 5 mm from the shell, **we estimate the time taken for coagulation, therefore for cooking, as** 2.7 minutes.

b) Hard-boiled egg

In the same way, using the modeling seen above in the albumin, looking for a temperature of 65°C at 1 cm from the shell, **we estimate the coagulation time, therefore the cooking time, to be** 17 minutes.

c) These results are relatively coherent with personal experience

Cooking a soft-boiled egg is of the order of 3.5 minutes. Modeling is correct here.

Cooking a hard-boiled egg is of the order of 10 minutes. This model gives a higher value.

Over and above the approximation made on the thermal properties of the white and the yolk, here, we use the results of a plane model in a revolving system. The geometry selected here is particularly schematic.

In addition, no influence has been attributed to the natural convection in the egg.

EXAMPLE 5.7.– Barbecue evening. Cooking meat on a plancha

A system of cooking on a plancha is made up of a rectangular heating plate which emits heat uniformly through its upper surface, of dimensions $50\,cm * 30cm$ and with a power of $P = 3600W$.

We place a piece of meat that is presumed to be perfectly plane onto the plancha. The heating surface is treated with a non-stick coating, and no fat or oil is required for cooking.

The useful thermal properties of the meat will be taken to be:

Thermal conductivity:

$$\lambda = 0.47\,Wm^{-2}K^{-2}$$

Density:

$$\rho = 1080\,kg.m^{-3}$$

Specific heat capacity:

$$c = 4000\,J\,kg^{-1}\,K^{-1}.$$

The slice of meat, which has just been taken out of the fridge, is at $T_i = 15°C$. We will consider that the meat will be cooked "rare" at a temperature of $T_{CS} = 50°C$.

We will also consider that, although the slice has a finite thickness, the evolution of temperatures within the meat is calculated as it is for a body of infinite thickness.

We also consider that the physical properties of the meat are modified very little by the cooking[1] (which is a debatable approximation):

1) We are seeking to determine the thickness of the meat cooked after a specific duration of time.

a) What is the density of the flow φ_0 imposed by the plate on the meat?

b) What thickness of meat will be cooked after 2 mn?

2) After these two minutes, what will the temperature T_S be at the surface of the meat placed on the plate? All good cooks know that meat must be "sealed" at the surface (meaning that the surface layer must be "pyrolyzed". Is this the case here?

3) A slice of meat with a thickness of 1.5 cm is cooked by placing it for two minutes each side on the plancha.

Will the meat be:

– cooked "rare"?

– cooked "medium rare"?

– grilled all the way through?

SOLUTION.–

1) We attempt to determine the thickness of the meat cooked after a determined amount of time.

a) The flow density φ_0 imposed by the plate on the meat is deduced from the flow, which is equal to the power dissipated by the plate:

$$\Phi = P = 3600\,W \qquad [5.188]$$

The surface of the plate is

$$S = \pi \frac{0.28^2}{4} = 6.16.10^{-2} \qquad [5.189]$$

1 We will admit it even for "pyrolyzed" meat at the surface (effect sought for "sealed" meat). This approximation is obviously a subject of discussion. We will not discuss this here.

$$\varphi_0 = \frac{\Phi}{S} = \frac{3600}{6.16.10^{-2}} = 2.4.10^4\, Wm^{-2} \qquad [5.190]$$

b) What thickness of meat will be cooked after 2 mn?

Let us firstly calculate the thermal diffusivity of the meat:

$$a = \frac{\lambda}{\rho c} = \frac{0.47}{1080 * 4000} = 1.09.10^{-7}\, ms^{-2} \qquad [5.191]$$

We will define the usual reduced variable:

$$\xi = \frac{x}{2\sqrt{at}} = \frac{x}{\sqrt{4at}} \qquad [5.192]$$

We need to resolve a problem of non-stationary conductive transfer with a flow density imposed on the input surface ($x = 0$). We need to calculate the temperature on this input face. To do so, we use the known result of the difference in temperature θ between the temperature at time t and the initial temperature:

$$\theta = T(x,t) - T_i \qquad [5.193]$$

$$\theta = \frac{2\varphi_0}{\lambda}\sqrt{at}\; ierfc(\xi) \qquad [5.194]$$

At time $t = 2\,mn = 120\,s$, we are looking for at what point the temperature is equal to $T_{CS} = 50°C$, in other words,

$$\theta = T(x,t) - T_i = 50 - 15 = 35°C \qquad [5.195a]$$

$$35 = \frac{2 * 2.4.10^4}{0.47}\sqrt{1.09.10^{-7} * 120}\; ierfc(\xi) \qquad [5.195b]$$

from this, we deduce:

$$ierfc(\xi) = \frac{35 * 0.47}{2 * 2.4.10^4 * \sqrt{1.09.10^{-7} * 120}} = 9.47.10^{-2} \qquad [5.196]$$

Hence,

$$\xi = 0.782 \quad [5.197]$$

$$\xi = 0.782 = \frac{x}{2\sqrt{1.09.10^{-7} * 120}} = \frac{x}{\sqrt{4at}} \quad [5.198]$$

$$x = 5.65.10^{-3}\ m = 5.65\ mm \quad [5.199]$$

2) What will the temperature T_S be after these two minutes, at the surface of the meat placed on the plate?

We look again at the solution giving $\theta(t)$

$$\theta = T_S - T_i \quad [5.200]$$

$$\theta = \frac{2\varphi_0}{\lambda}\sqrt{at}\ ierfc(\xi) \quad [5.201]$$

$$\theta = \frac{2 * 2.4.10^4}{0.47}\sqrt{1.09.10^{-7} * t}\ ierfc(\xi) \quad [5.202]$$

$$\theta = 33.7\sqrt{t}\ ierfc(\xi) \quad [5.203]$$

$$\xi = \frac{x}{\sqrt{4at}} = \frac{x}{6.6.10^{-4} * \sqrt{t}} = 0 \quad [5.204]$$

$$ierfc(0) = 0.5642 \quad [5.205]$$

$t = 2\,mn = 120\,s$ [2] [5.187]

$$\theta = 33.7 * \sqrt{120} * 0.5642 = 208°C \quad [5.206]$$

$$T_S = 233°C \quad [5.207]$$

And of course: the meat will be "sealed".

3) We cook a slice of meat of thickness $e = 1.5\,cm$ by placing it on the plancha for two minutes on each side.

The penetration distance at 50°C is 5.65 mm.

We will have "rare" cooked meat for $2 * 0.565 = 1.13\,cm$ and the center will be very "raw" for $1.5 - 1.13 = 0.37cm = 3.7\ mm$, with a well-sealed surface.

The meat will be rare and cooked well.

EXAMPLE 5.8.– Fusion of a weld

A cube with sides $a = 10\,cm$ is made up of two parallelepipeds A and B, each with sides $a * a * \frac{a}{2}$ welded with tin on its large faces.

We estimate that the welding is a very thin film of tin that does not act on the temperature distribution in the copper block with sides a.

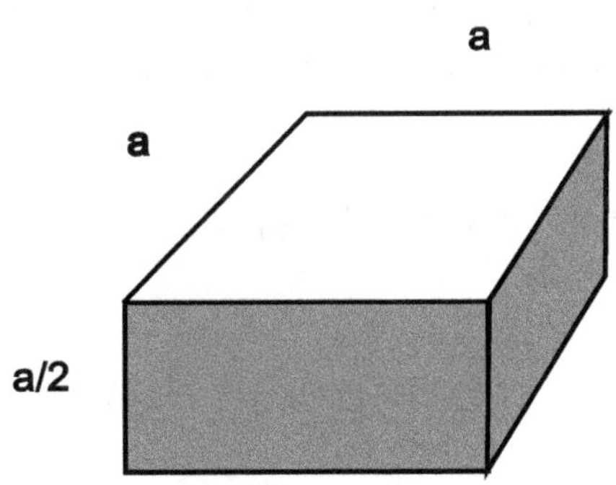

Figure. 5.9. *A block for welding*

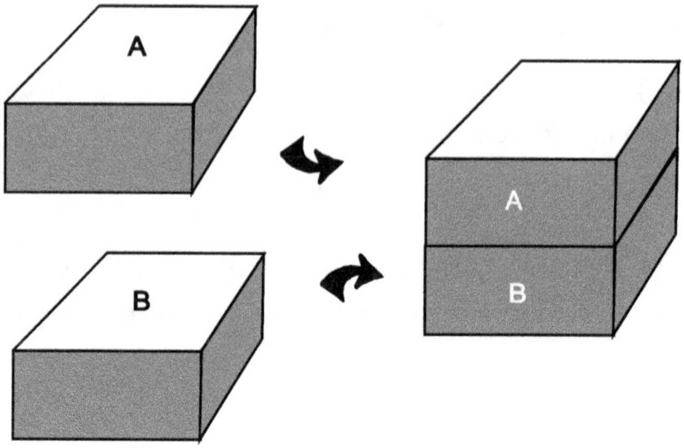

Figure 5.10. *The welding*

The copper cube is balanced with the atmospheric temperature $T_i = 21°C$.

We fix the copper cube below a horizontal plate P at the temperature of 310°C.

The fixing is sufficiently rapid for us to estimate that the upper face of the cube is **rapidly brought to a temperature of** $T_0 = 310°C$**, which it will retain.**

We give the thermal properties of copper:

Density:

$$\rho = 8954\ kg.m^{-3}$$

Thermal conductibility:

$$\lambda = 386\ kg.m^{-3}\ W\ m^{-1}\ K^{-1}$$

Specific heat capacity:

$$c = 383\ J.kg^{-1}.$$

We presume that the transfer of non-stationary heat in the copper cube is practically single-dimensional.

So, Ox is the vertical axis, with origin O at the center of the upper face of the cube.

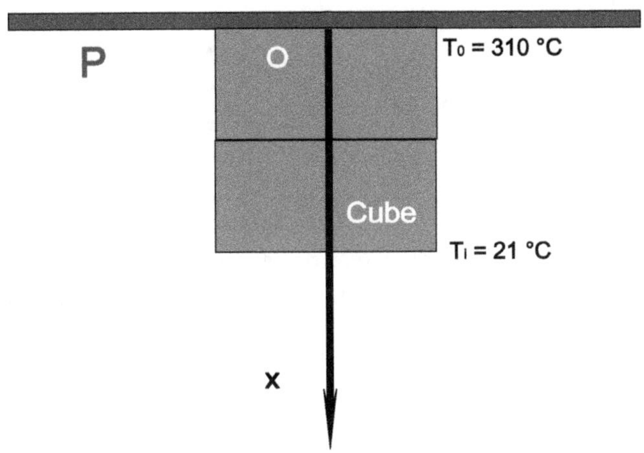

Figure 5.11. *Diagram of the device*

1) Give the general expression for the temperature $T(x,t)$ of any horizontal plane in the cube.

2) Tin melts at 232°C. How long after the cube and the plate are put into contact will the cube B fall?

SOLUTION.–

1) We are in the presence of a non-stationary conductive transfer at an imposed wall temperature. The solution is known:

Defining:

$$\theta = \frac{T(x,t) - T_e}{T_i - T_e} \qquad [5.208]$$

we have:

$$\theta = erf\left(\xi\right) \qquad [5.209]$$

where:

$$\xi = \frac{x}{\sqrt{4at}} \qquad [5.210]$$

Writing $T\left(x,t\right)$ explicitly: [5.211]

$$T\left(x,t\right) = \left(T_i - T_e\right) erf\left(\xi\right) + T_e \qquad [5.212]$$

2) **The tube will fall when the tin has melted**. This will intervene at time t_F where the temperature will reach $T_F = 232\,°C$ in the plane that separates the two half-cubes, in other words at $x = \frac{a}{2} = 5\,cm = 5.10^{-2}\ m$.

We calculate the thermal diffusivity of the copper:

$$a = \frac{\lambda}{\rho c} = \frac{386}{8954 * 383} = 1.126.10^{-4}\ m^2.s^{-1} \qquad [5.213]$$

The desired value of θ is:

in other words:

$$\theta = \frac{232 - 310}{21 - 310} = 0.27 \qquad [5.214]$$

Consulting the table for $erf\left(\xi\right)$, by interpolation, we find that:

$$0.27 = erf\left(0.244\right) \qquad [5.215]$$

From this, we deduce the time that we are looking for

$$\xi\left(x,t\right) = \frac{x}{2\sqrt{at}} = 0.244 = \frac{5.10^{-2}}{\sqrt{4 * 1.126.10^{-4} * t}} \qquad [5.216]$$

$$t = \left(\frac{5.10^{-2}}{0.244} \right)^2 \frac{1}{4*1.126.10^{-4}} \qquad [5.217]$$

$$t = 93.23\ s \qquad [5.218]$$

EXAMPLE 5.9.– Welding on copper

We wish to weld a plate of copper of thickness $e = 5\ mm$ onto a base that has already been covered with a thin layer of tin. Initially, the plate, tin and its base are all at a homogeneous temperature of $T_i = 21°C$.

For the plate of zinc, we place it on the layer of tin. We presume that the contact is perfect and that no layer of air can pass between the copper and the tin.

On the upper part of the copper, we place a soldering iron tip with a surface that is presumed to be much larger than the soldering zone. The problem of conduction will thus remain purely single-dimensional.

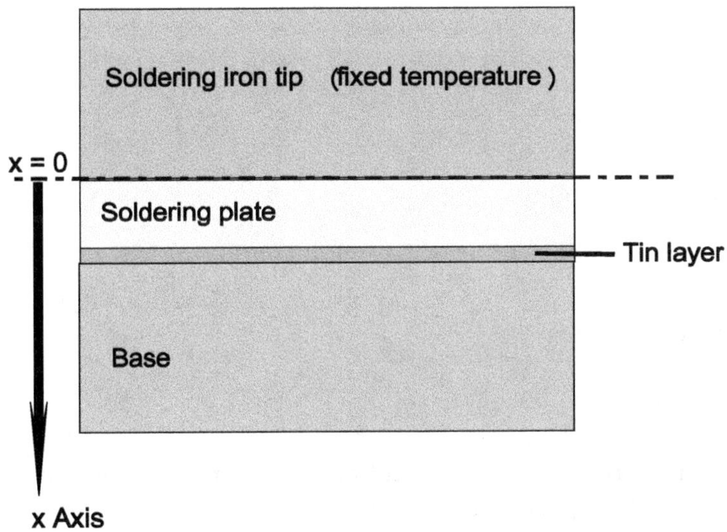

Figure 5.12. *Diagram of a tin weld*

We will try to approach the problem with two different hypotheses (questions **1** and **2**)

We give:

For the copper

Density:

$$\rho = 8954\,kg.m^{-3}$$

Thermal conductibility:

$$\lambda = 386\,Wm^{-2}K^{-2}$$

Specific heat capacity:

$$c = 383\,J.kg^{-1}$$

For the zinc

Density:

$$\rho = 7144\,kg.m^{-3}$$

Thermal conductibility:

$$\lambda = 112\,Wm^{-2}K^{-2}$$

Specific heat capacity:

$$c = 384\,J.kg^{-1}$$

1) We presume that the welding iron imposes a temperature $T_e = 400\,°C$ on the upper surface of the copper plate.

a) The tin melts at $T_F = 232\,°C$. What is the minimum time that the iron tip must be left on the plate of copper so that the weld can be performed?

b) Look again at the problem for a zinc plate of thickness 3 mm.

2) We look again at the problem of the copper plate of 5 mm in another hypothesis. **We will presume, this time, that the welding iron imposes a thermal flow density of $\varphi = 5.10^5\ W.m^{-2}$ on the upper surface of the plate.**

After time t found in **1a**, what will the temperature be in the tin?

What conclusions do you draw from this concerning the validity of our previous calculation? In order to reason this answer as best possible, it will be useful for us to be able to calculate the temperature of the upper surface of the copper plate at time t calculated in **1a**.

SOLUTION.–

1) We have a non-stationary transfer problem at an imposed temperature.

a) We will calculate the thermal diffusivity of the copper:

$$a = \frac{\lambda}{\rho c} = \frac{386}{8954 * 383} = 1.125.10^{-4}\ m^2 s^{-1} \quad [5.219]$$

We will define a reduced variable:

$$\theta = \frac{T - T_e}{T_i - T_e} \quad [5.220]$$

with

$$T_i = 21°C \quad [5.221a]$$

$$T_e = 400°C \quad [5.221b]$$

We know that θ will be of the form

$$\theta = erf\ \xi \quad [5.222]$$

with:

$$\xi = \frac{x}{\sqrt{4at}} \quad [5.223]$$

We want to reach

$$T\left(x,t\right)=232°C \qquad [5.224]$$

at the distance from the iron tip equal to $x=e=5\ mm$

We have

$$\theta=0.443 \qquad [5.225]$$

and the table gives us, following interpolation:

$$\xi=0.415 \qquad [5.226]$$

from this, we then deduce:

$$0.415=\frac{5.10^{-3}}{\sqrt{4*1.125.10^{-4}\,t}}=\frac{0.235}{\sqrt{t}} \qquad [5.227]$$

$$\sqrt{t}=0.566 \qquad [5.228]$$

$$t=0.32\,s \qquad [5.229]$$

b) With the zinc plate, we take the same approach

$$a=\frac{\lambda}{\rho c}=\frac{112}{7144*384}=4.08.10^{-5}\ m^2s^{-1} \qquad [5.230]$$

We will define a reduced variable:

$$\theta=\frac{T-T_e}{T_i-T_e} \qquad [5.231]$$

with

$$T_i=21°C \qquad [5.232]$$

$$T_e=400°C \qquad [5.233]$$

We know that θ will be of the form

$$\theta = erf\ \xi \quad [5.234]$$

with:

$$\xi = \frac{x}{\sqrt{4at}} \quad [5.235]$$

We wish to reach

$$T(x,t) = 232°C \quad [5.236]$$

at the distance from the iron tip equal to

$$x = e = 5\ mm \quad [5.237]$$

We have

$$\theta = 0.443 \quad [5.238]$$

and the table gives us, after interpolation:

$$\xi = 0.415 \quad [5.239]$$

from this, we then deduce:

$$0.415 = \frac{3.10^{-3}}{\sqrt{4 * 4.08.10^{-5}\ t}} = \frac{0.235}{\sqrt{t}} \quad [5.240]$$

$$\sqrt{t} = 0.566 \quad [5.241]$$

$$t = 0.32\ s \quad [5.242]$$

We obtain the same time for this thinner zinc plate.

2) In the context of this second hypothesis, we have a problem of an imposed flow density

$$\theta = T(x,t) - T_i \quad [5.243]$$

We know that the difference of temperature between $T(x,t)$ and the reduced temperature obeys

$$\theta = \frac{2\varphi_0}{\lambda}\sqrt{at}\ ierfc(\xi) \quad [5.244]$$

We are looking for the temperature at $x = 5\,mm$ and $t = 0.32\,s$

ξ will have the value previously seen of $\xi = 0.415$ [5.245]

The corresponding value of $ierf(\xi)$ is given by the table:

$$ierf(\xi) = 0.2429 \quad [5.246]$$

We therefore find

$$\theta = 3.77°C \quad [5.247]$$

which leads to

$$T = 24.77°C \quad [5.248]$$

which is obviously highly insufficient for a good weld. The welding time has been significantly underestimated by the first hypothesis, which must therefore be pushed back in time.

We can understand this by calculating the temperature at the point of contact of the iron tip, so, for $x = 0$

$$\theta(0, 0.32) = \frac{2\varphi_0}{\lambda}\sqrt{at}\ ierfc(0) = 8.49°C \quad [5.249]$$

The temperature on contact with the welding iron is

$$T = 29.49°C \quad [5.250]$$

6

Fin Theory: Notions and Examples

6.1. Notions regarding the theory of fins

6.1.1. *Principle of fins*

In principle, the use of a fin to dissipate heat is based on a simple idea.

Let us look at an example. A cylinder with radius $r = 5mm$ and height $h = 3cm$ generates an internal power of $P = 3W$, which will be equal to the flow Φ_0 that must be dissipated. Its temperature must not exceed T_0 = 85°C. We suppose an ambient temperature of T_e = 20°C.

If we assume natural convection is cooling this cylinder, for example with a convection coefficient $h = 10W.m^2.K^{-1}$, the exchange surface will be $S = 9.4.10^{-4} m^2$ and the temperature of the cylinder will be such that $P = \Phi_0 = S\, h(T_{01} - T_e)$; in other words, T_{01} = 339°C, which is much higher than T_0 = 70°C. We can consider increasing the convection coefficient that should then be equal to $h' = 63.8\, W.m^2.K^{-1}$. Perhaps this could be imagined for forced convection, but at the price of a significant technical complication.

We can more simply imagine welding eight parallelepiped plates, or wings, on the initial cylinder, with a cross-section of $0.3 * 3$ cm² and length of $L = 1cm$. Neglecting the surface of the body that has remained free (the wings occupy 2.4 cm of 3.14 cm of the outside), the exchange surface of the eight wings will become $S' = 8 * 6.6.10^{-2} * 1.10^{-2} = 5.28.10^{-3}$ m^2.

We will choose, for example, aluminum to construct these fins. Supposing that the plate conducts heat infinitely, the surface of the fin will be uniform and equal to that of the cylinder that requires cooling, in other words T_{02}. According to $P = \Phi_0 = S\,h(T_{02} - T_e)$, this new temperature is denoted as $T_{02} = 77°\text{C}$, which now fulfills the condition $T_{02} < T_0$.

However, in reality, things do not happen like this.

a) The fin is not an infinite conductor, and a longitudinal temperature gradient is required.

b) Consequently, the flow evacuated by the sides of the wing reduces the cylinder towards the free end of the wing. The flow evacuated in reality Φ will therefore necessarily be lower than the "ideal" flow found for an isothermal wing.

Furthermore, we will see that the relation between the two flows defines the efficiency η of the fin:

$$\eta = \frac{\Phi}{\Phi_0}$$

6.1.2. *Elementary fin theory*

A wing takes the form of an object, generally metallic, with one end welded to a body that has a temperature T_0. This object will have a cross-section S that can vary as a function of the distance from one of its cross-sections with respect to the body.

Two main geometries can exist: a linear fin and a circular fin.

The temperature T_0 will constitute a limit condition in this problem. It may be that it is determined in the body to be cooled by other phenomena, in particular the generation of a quantity of heat in the body.

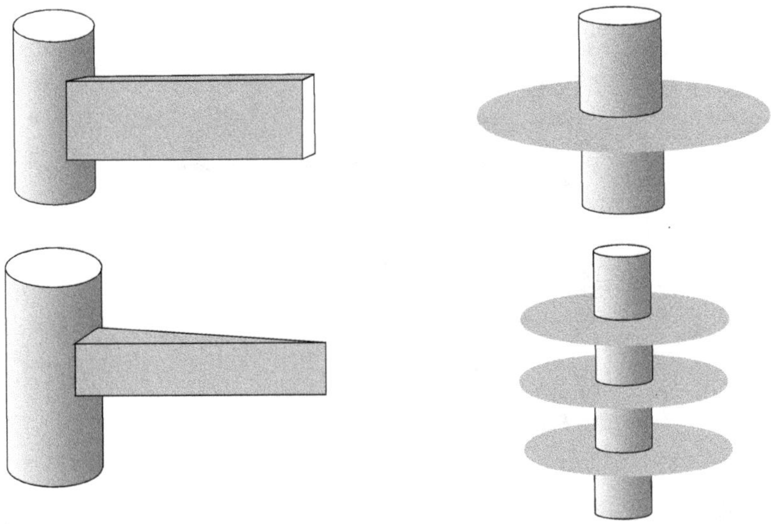

Linear fins Cylindrical, simple or multiple fins

Figure 6.1. *Two types of fin: linear and cylindrical*

The basic problem that we have to resolve each time we study a wing is as follows.

Given the temperature T_0 of the body to which a fin is welded and the ambient temperature T_e, what is the total flow of heat that escapes from the fin? This flow will correspond to the quantity of heat removed from the body that needs to be cooled.

To resolve this problem, we will divide the fin into infinitely thin sections, and write a thermal balance for each of these sections:

Flow entering by conduction = Flow exiting by conduction + Flow exiting by convection.

We thus obtain a differential equation (a special form of the general equation of stationary thermal conduction) that will be resolved by applying the limit conditions, which will vary from one problem to another.

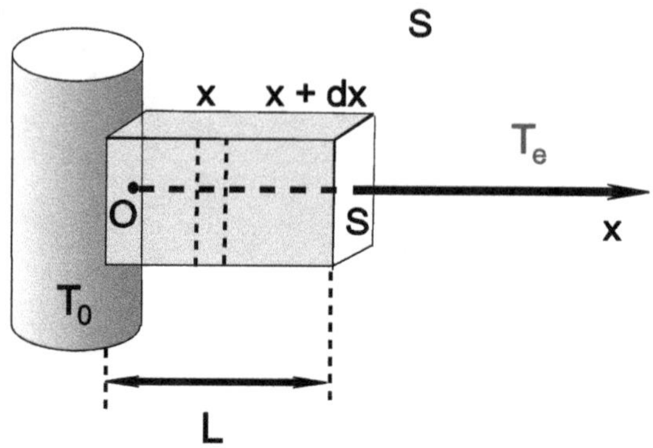

Figure 6.2. *Model of the parallelepiped linear fin*

We will locate the various sections of the fin on the Ox axis that is perpendicular to the body considered, where the origin O is on the body in question. At this stage of the equation, the fin is not a parallelepiped, and therefore $S = S(x)$.

We will set out a fundamental hypothesis, which is common when dealing with these problems: the field of temperatures will be single-dimensional. In other words, the isotherms of the fin are perpendicular planes to the Ox axis; or $T = T(x)$.

This hypothesis presumes that the lateral thermal gradients are negligible in comparison with the longitudinal gradients. Physically, this comes down to assuming that the lateral convective flows are weak in comparison with the axial conductive flow.

In addition, the fin will be made up of a homogeneous material of constant conductibility λ.

On the lateral surface of the wing, the convective transfer is determined by a convection coefficient h. We will require an additional local parameter; the perimeter of the section for which we will write the balance as: $p = p(x)$.

We will consider a section of the fin of thickness dx located at a distance x from the body.

We will write the flow. First, we note that the flows are directed from the body towards the end of the wing; therefore, the temperature decreases according to Ox and $\frac{dT}{dx} < 0$. The flow entering by conduction is given by:

$$\Phi_e = -\lambda S(x)\frac{dT}{dx} \tag{6.1}$$

This quantity is positive.

The flow exiting by conduction is given by:

$$\Phi_S = -\lambda S(x)\frac{dT}{dx} + \frac{d}{dx}\left(-\lambda S(x)\frac{dT}{dx}\right)dx \tag{6.2}$$

This quantity is also positive.

The flow exiting by convection:

$$\Phi_{Conv} = \left[p(x)dx\right]h\left(T(x) - T_e\right) \tag{6.3}$$

In this expression, $p(x)dx$ is the lateral surface through which the flow taken out by convection flows.

The incoming flow is divided into exiting conductive flow and convective flow to the outside. The balance is therefore written as:

$$\Phi_e = \Phi_S + \Phi_{Conv} \tag{6.4}$$

$$\begin{aligned} -\lambda S(x)\frac{dT}{dx} &= -\lambda S(x)\frac{dT}{dx} + \frac{d}{dx}\left(-\lambda S(x)\frac{dT}{dx}\right)dx \\ &+ \left[p(x)dx\right]h\left(T(x) - T_e\right) \end{aligned} \tag{6.5}$$

After simplification:

$$\frac{d}{dx}\left(\lambda S(x)\frac{dT}{dx}\right) - hp(x)\left(T(x) - T_e\right) = 0 \tag{6.6}$$

We note that in the most general problem, the term in $S(x)$ remains in the derivative, which can lead to equations that are complex to resolve.

6.1.3. *Parallelepiped fin*

6.1.3.1. *Solution*

To illustrate the type of solutions found, we consider here the simplest case, that of a parallelepiped fin. We also provide a few other solutions in Appendix 6.

First, a point about vocabulary and notation must be specified. We call S the face of the wing that is parallel to the surface of the body (at the extremity origin ($x = 0$)); a surface S is fixed to the body to be cooled.

In the case of the parallelepiped wing, S is a rectangle with sides l and e. We will call the largest of these sides the width l of the wing, and the smallest of these sides the thickness e. The third dimension of the parallelepiped will be defined as the length L of the wing. It should be noted that for a short wing, the "length" of the wing can be smaller than its "width". For example, we can have: $l = 3\,cm,\, e = 3\,m,\, L = 1\,cm$.

Then, for this parallelepiped wing:

$$S = l\,e = Cte \qquad [6.7]$$

and

$$p = 2(l + e) = Cte \qquad [6.8]$$

The balance equation will therefore simplify to:

$$\lambda S \frac{d^2T}{dx^2} - h\,p\,(T(x) - T_e) = 0 \qquad [6.9]$$

where, again:

$$\frac{d^2T}{dx^2} - \omega^2\,(T(x) - T_e) = 0 \qquad [6.10]$$

Here, this has led to the introduction of a parameter ω, a classic for these fin problems. This parameter is defined by:

$$\omega^2 = \frac{h\,p}{\lambda S} \qquad [6.11]$$

For ease of calculation, we will introduce the reduced variable:

$$\theta(x) = T(x) - T_e \qquad [6.12]$$

Then:

$$\frac{d\theta(x)}{dx} = \frac{dT(x)}{dx} \qquad [6.13]$$

The equation to resolve becomes:

$$\frac{d^2\theta}{dx^2} - \omega^2\,\theta = 0 \qquad [6.14]$$

Various conditions can then be envisaged:

a) The fin is infinitely long. In terms of limit conditions, we therefore have:

$$x = 0 \quad ; \quad \theta = T_0 - T_e = \theta_0 \qquad [6.15]$$

$$x \to \infty \quad ; \quad \theta \to T_e - T_e = 0 \qquad [6.16]$$

b) The fin has a finite length of $T_0 - T_e = \theta_0$. We assume that the dissipated flow at its extremity is negligible. We therefore have the limit conditions:

$$x = 0 \quad ; \quad \theta = T_0 - T_e = \theta_0 \qquad [6.17]$$

$$x = L \quad ; \quad -\lambda\frac{dT}{dx} = 0 \Rightarrow \frac{d\theta}{dx} = 0 \qquad [6.18]$$

Here, we will look at these two geometries.

We can also take into account the flow at the extremity. Keeping the same convection coefficient to simplify, we will have:

$$x = 0 \quad ; \quad \theta = T_0 - T_e = \theta_0 \qquad [6.19]$$

$$x = L \quad ; \quad -\lambda \frac{dT}{dx} = h\,(T - T)_e \Rightarrow \frac{d\theta}{dx} = -\frac{h}{\lambda}\theta \qquad [6.20]$$

The general solution to the differential equation

$$\frac{d^2\theta}{dx^2} - \omega^2\,\theta = 0 \qquad [6.21]$$

is known.

We can write it in two forms:

the most common one is:

$$\theta = Ae^{\omega x} + Be^{-\omega x} \qquad [6.22]$$

or a form that uses hyperbolic functions:

$$\theta = A\,sh\,\omega x + B\,ch\,\omega x \qquad [6.23]$$

This last form will most often be preferred for practical reasons.

Moreover, these two expressions are equivalent, since the sine and cosine hyperbolics $sh\,\omega x$ and $ch\,\omega x$ are linear forms of the exponentials $e^{\omega x}$ and $e^{-\omega x}$.

NOTE.– Readers who are not familiar with hyperbolic functions can refer to Appendix 8, which contains a few reminders that are essential for these functions.

We will distinguish two cases, as indicated above:

– the wing is long enough to be assimilated to an infinite length;

– the wing has a finite length.

6.1.3.2. *Problem of an infinite fin*

We need to resolve the equation:

$$\frac{d^2\theta}{dx^2} - \omega^2\theta = 0 \qquad [6.24]$$

with the limit conditions:

$$x = 0 \quad ; \quad \theta = T_0 - T_e = \theta_0 \qquad [6.25]$$

$$x \to \infty \quad ; \quad \theta \to T_e - T_e = 0 \qquad [6.26]$$

In this specific case, we will use the solution of the sum of two exponentials:

$$\theta = Ae^{\omega x} + Be^{-\omega x} \qquad [6.27]$$

The two constants are determined by the limit conditions:

$$x = 0 \quad ; \quad \theta = \theta_0 = Ae^0 + Be^0 = A + B \qquad [6.28]$$

$$x \to \infty \quad ; \quad \theta = Ae^{\infty} + Be^{-\infty} \to 0 \qquad [6.29]$$

where A placed in front of an infinite term is necessarily zero. The following holds:

$$B = \theta_0 \qquad [6.30]$$

We therefore obtain:

$$\theta = \theta_0 e^{-\omega x} \qquad [6.31]$$

and then:

$$T(x) = (T_0 - T_e)e^{-\omega x} + T_e \qquad [6.32]$$

The parameter of interest in this type of problem is the flow Φ that is evacuated, which will be the flow observed at the connection between the wing and the body, that is, at $x = 0$:

$$\Phi = -S\lambda \left.\frac{dT}{dx}\right|_{x=0} = -S\lambda(T_0 - T_e)\left(-\omega e^{-\omega x}\right)\Big|_{x=0} \qquad [6.33]$$

$$\Phi = -S\lambda \left.\frac{dT}{dx}\right|_{x=0} = -S\lambda(T_0 - T_e)\left(-\omega e^{-\omega x}\right)\Big|_{x=0} \qquad [6.34]$$

$$\Phi = S\lambda\omega(T_0 - T_e) = \theta_0 S\lambda\omega \qquad [6.35]$$

which can then be written by the definition of:

$$\omega = \sqrt{\frac{hp}{\lambda S}} \qquad [6.36]$$

$$\Phi = S\lambda\omega(T_0 - T_e) = S\lambda\sqrt{\frac{hp}{\lambda S}}(T_0 - T_e) \qquad [6.37]$$

$$\Phi = \sqrt{S\lambda\, hp}(T_0 - T_e) = \theta_0\sqrt{S\lambda\, hp} \qquad [6.38]$$

We can compare this flow to Φ_0, the flow that we would have with an ideal wing; in other words, the isotherm at the temperature of the body needs to be cooled. This flow Φ_0 is calculated by considering:

$$\Phi_0 = pLh(T_0 - T_e) = pLh\theta_0 \qquad [6.39]$$

6.1.3.3. *Problem of a fin of finite length*

We now consider a fin of the same geometry as previously examined, with a length that is finite and equal to L. We will neglect the thermal flow at the end of the fin.

We need to resolve the equation:

$$\frac{d^2\theta}{dx^2} - \omega^2\,\theta = 0 \qquad [6.40]$$

with the limit conditions:

$$x = 0 \quad ; \quad \theta = T_0 - T_e = \theta_0 \qquad [6.41]$$

$$x = L \quad ; \quad -\lambda\frac{dT}{dx} = 0 \Rightarrow \frac{d\theta}{dx} = 0 \qquad [6.42]$$

Here, we will use hyperbolic functions to write the general solution to this equation:

$$\theta = A\,sh\,\omega x + B\,ch\,\omega x \qquad [6.43]$$

The two constants are determined by the limit conditions:

$$x = 0 \quad ; \quad \theta = T_0 - T_e = \theta_0 = A\,sh0 + B\,ch0 = B \qquad [6.44]$$

$$x = L \quad ; \quad -\lambda\frac{dT}{dx} = 0 \Rightarrow \frac{d\theta}{dx} = A\omega ch\,\omega L + B\,\omega\,sh\,\omega L = 0 \qquad [6.45]$$

We obtain:

$$B = \theta_0 \qquad [6.46]$$

We therefore obtain:

$$A = -\frac{B\,\omega\,sh\,\omega L}{\omega ch\,\omega L} = -\frac{\theta_0\ sh\,\omega L}{ch\,\omega L} = -\theta_0\ th\,\omega L \qquad [6.47]$$

and then:

$$\theta(x) = \theta_0\left(\frac{-\ sh\omega L}{ch\omega L}sh\omega x + ch\omega x\right) \qquad [6.48]$$

After a few transformations:

$$\theta(x) = \theta_0 \left(\frac{-\,sh\,\omega L\, sh\,\omega x + ch\,\omega L\, ch\,\omega x}{ch\,\omega L} \right)$$
$$= \theta_0 \left(\frac{-\,sh\,\omega L\, sh\,\omega x + ch\,\omega L\, ch\,\omega x}{ch\,\omega L} \right) \qquad [6.49]$$

We know that:

$$ch(x-y) = ch\,x\,ch\,y - sh\,x\,sh\,y \qquad [6.50]$$

$\theta(x)$ can be rewritten as:

$$\theta(x) = \ = \theta_0 \left(\frac{-\,sh\omega L\, sh\omega x + ch\omega L\, ch\omega x}{ch\omega L} \right) = \theta_0 \left(\frac{ch\omega(L-x)}{ch\omega L} \right) \qquad [6.51]$$

Or even:

$$T(x) = \theta_0 \left(\frac{ch\omega(L-x)}{ch\omega L} \right) + T_e = (T_0 - T_e)\left(\frac{ch\omega(L-x)}{ch\omega L} \right) + T_e \qquad [6.52]$$

The parameter of interest in this type of problem is the evacuated flow Φ that will be the observed flow at the connection between the wing and the body, so at $x = 0$:

$$\Phi = -S\lambda \left.\frac{dT}{dx}\right|_{x=0} = -S\lambda (T_0 - T_e)\left(\frac{-\omega sh\omega(L-x)}{ch\omega L} \right) + T_e \Bigg|_{x=0} \qquad [6.53]$$

$$\Phi = S\lambda (T_0 - T_e)\left(\frac{\omega sh\omega L}{ch\omega L} \right) = S\lambda\omega (T_0 - T_e)\, th\,\omega L \qquad [6.54]$$

which can be written by the definition of:

$$\omega = \sqrt{\frac{hp}{\lambda S}} \qquad [6.55]$$

$$\Phi = S\lambda\omega(T_0 - T_e)th\omega L = S\lambda\sqrt{\frac{hp}{\lambda S}}(T_0 - T_e)th\omega L \quad [6.56]$$

$$\Phi = \sqrt{S\lambda\, hp}(T_0 - T_e)th\omega L = \theta_0\sqrt{S\lambda\, hp}\, th\omega L \quad [6.57]$$

6.2. Examples of application

In the above fin theory, the transfers involved are reduced to conduction and convection. A wing can also have radiation. Examples of this case have been attributed to the section dedicated to radiation in **Chapter 5**.

EXAMPLE 6.1.– Rectangular fin

A rectangular wing is mounted on a body at constant temperature $T_0 = 75\,°C$. The ambient temperature is $T_e = 25°C$.

The convection coefficient between the fin and the ambient air is $h = 10W.m^{-2}.K^{-1}$.

The surface area of the fin (not including the part welded to the body to be cooled) is $S = 20\,cm^2$.

1) If the entire surface of the wing was at T_0, what would the flow Φ_0 evacuated by the wing be?

2) The flow evacuated by the wing is, in fact, $\Phi = 0.9W$. What is the efficiency η of the fin?

SOLUTION.–

1) The flow that will escape by convection of an isothermal wing at $T(x) \equiv T_0$ over the entire surface area S can be simply written as:

$$\Phi_0 = Sh(T_0 - T_e) \quad [6.58]$$

$$\Phi_0 = 20.10^{-4}\,10(75 - 25) = 1W \quad [6.59]$$

2) By definition of the efficiency:

$$\eta = \frac{\Phi}{\Phi_0} = 0.9 \qquad [6.60]$$

EXAMPLE 6.2.– Calculation of a fin

An aluminum fin of width $l = 5cm$, length $L = 10cm$ and thickness $e = 3mm$ is embedded in a wall. The conductibility of the aluminum is $\lambda = 204 W.m^{-1}.K^{-1}$.

The base of the wall is maintained at $T_0 = 300\,°C$, the ambient temperature is $T_e = 30\,°C$ and the coefficient of convective transfer is $h = 10 W.m^{-2}.K^{-1}$.

1) Determine the temperature T_F at the extremity of the fin and the flow Φ extracted by the fin if we neglect the thermal gradients in the width and thickness directions.

2) The efficiency η of the fin is the ratio of the flow extracted Φ, in comparison with the flow Φ_0 that would be extracted by the wing of the same geometry, whose temperature will be uniform and equal to the temperature of its base. Calculate this efficiency.

3) Calculate the relative error that would be made by considering the temperature of the end of the fin is equal to the ambient temperature.

SOLUTION.–

1) We define

The reduced variable:

$$\theta = T(x) - T_e \qquad [6.61]$$

$$\theta_0 = T_0 - T_e \qquad [6.62]$$

The perimeter p_e of the fin is:

$$p_e = 2(l + e) \qquad [6.63]$$

The parameter ω is defined as:

$$\omega^2 = \frac{h_e\, p_e}{\lambda S} \qquad [6.64]$$

For a wing of finite length:

$$\Phi = \lambda S\, \omega \theta_0\, th(\,\omega L) = \sqrt{h\, p_e\, \lambda S}\;\theta_0\, th(\,\omega L) \qquad [6.65]$$

For an infinitely long fin, we will have:

$$\Phi = \lambda S\, \omega \theta_0 = \sqrt{h\, p_e\, \lambda S}\;\theta_0 \qquad [6.66]$$

We obtain:

$$p_e = 2(l + e) = 2\left(5.10^{-2} + 3.10^{-3}\right) = 0.106\, m \qquad [6.67]$$

$$S = l e = 5.10^{-2} * 3.10^{-3} = 15.10^{-5}\, m^2 \qquad [6.68]$$

$$\omega = \sqrt{\frac{10 * 0{,}106}{204 * 1.5.10^{-4}}} = \sqrt{34.64} = 5.88\, m^{-1} \qquad [6.69]$$

$$\Phi = \sqrt{h\, p_e\, \lambda S}\;\theta_0\, th(\,\omega L)$$

$$= \sqrt{10 * 0.106 * 204 * 1.5.10^{-4}}\;(300 - 30)\, th\,(0.588) = 25.72\, W$$

$$th\,(0.588) = \frac{1.8004 - 0.555}{1.8004 + 0.555} = 0.529 \qquad [6.70]$$

2) Efficiency of the fin

The flow that would escape by convection of an isothermal fin at $T(x) \equiv T_0$ over the full lateral surface area S_S can be simply written as:

$$S_S = p_e\, L \qquad [6.71]$$

$$\Phi_0 = S_S\, h(T_0 - T)_e = p_e\, L h(T_0 - T)_e = p_e\, L h \theta_0 \qquad [6.72]$$

So, the general expression for the ratio:

$$\eta = \frac{\Phi}{\Phi_0} = \frac{\sqrt{h\, p_e\, \lambda S}\ \theta_0\, th(\,\omega L)}{p_e\, L h \theta_0} = \frac{\sqrt{(h\, p_e)^2 \omega^2}\ \theta_0\, th(\,\omega L)}{p_e\, L h \theta_0} = \frac{th(\,\omega L)}{\omega L} \qquad [6.73]$$

$$\eta = \frac{th(\,\omega L)}{\omega L} = \frac{0.529}{0.588} = 0.899 \approx 0.9 \qquad [6.74]$$

3) The balance equation for the fin must be resolved with the imposed limit condition. In other words:

$$\lambda S \frac{d^2 T}{d\, x^2} + h\, p_e (T - T_e) = 0 \qquad [6.75]$$

Introducing the variable:

$$\theta = T(x) - T_e \qquad [6.76]$$

$$\theta_0 = T_0 - T_e \qquad [6.77]$$

and the parameter ω such that:

$$\omega^2 = \frac{h_e\, p_e}{\lambda S} \qquad [6.78]$$

The equation becomes:

$$\frac{d^2 \theta}{d\, x^2} + \omega^2 \theta = 0 \qquad [6.79]$$

The two limit conditions become:

$$x = 0 \quad ; \quad \theta = T_0 - T_e = \theta_0 \qquad [6.80]$$

$$x = L \quad ; \quad \theta = T_e - T_e = 0 \qquad [6.81]$$

The general solution can be expressed by a sum of sinusoidal functions or hyperbolic functions. The second way of writing it is more suitable:

$$\theta = A\, sh\,\omega x + B\, ch\,\omega x \qquad [6.82]$$

The constants A and B are determined by the limit conditions:

$$x = 0 \quad ; \quad \theta = \theta_0 = A\, sh 0 + B\, ch 0 = B \qquad [6.83]$$

$$x = L \quad ; \quad \theta = 0 = A\, sh\,\omega L + B\, ch\,\omega L \qquad [6.84]$$

$$B = \theta_0 \qquad [6.85]$$

$$A\, sh\,\omega L = -B\, ch\,\omega L \qquad [6.86]$$

$$A = -\frac{\theta_0}{th\,\omega L} \qquad [6.87]$$

For the expression $\theta(x)$ and the flow Φ exiting the wing (calculated at $x = 0$), we have:

$$\frac{\theta}{\theta_0} = -\frac{1}{th\,\omega L} sh\,\omega x + ch\,\omega x \qquad [6.88]$$

$$\begin{aligned} \Phi &= -S\lambda \left.\frac{dT}{dx}\right|_{x=0} = -S\lambda \left.\frac{d\theta}{dx}\right|_{x=0} \\ &= -S\lambda \frac{d}{dx}\left[\theta_0\left(-\frac{\omega}{th\,\omega L} sh\,\omega x + ch\,\omega x\right)\right]\Bigg|_{x=0} \end{aligned} \qquad [6.89]$$

$$\Phi = -S\lambda\theta_0 \frac{\omega(-ch\omega x + sh\omega x)}{th\omega L}\Bigg|_{x=0} = -S\lambda\theta_0 \frac{-\omega ch(0)}{th\omega L} \qquad [6.90]$$

$$\Phi = S\lambda\theta_0\,\omega \frac{1}{th\omega L} = \sqrt{S\lambda h\,p_e}\,\theta_0 \frac{1}{th\omega L} \qquad [6.91]$$

$$\Phi = \frac{\sqrt{h\,p_e\,\lambda S}\;\theta_0}{th(\,\omega L)} = \frac{\sqrt{10*0,106*204*1,5.10^{-4}}\;(300-30)}{th\,(0,588)}t = 90,92\,W \qquad [6.92]$$

with:

$$th\,(0.588) = 0.529 \qquad [6.93]$$

We overestimate the evacuated flow of:

$$\frac{90.92-25.72}{25.72} = 253\% \qquad [6.94]$$

This is explained by the fact that the actual temperature found, presuming that the flow at the end of the wing is zero, can be written as:

$$\frac{\theta}{\theta_0} = -th\omega L\,sh\omega x + ch\omega x \qquad [6.95]$$

$$\theta(L) = -\theta_0\,(th\omega L\,sh\omega L + ch\omega x) = \theta_0 \frac{ch^2\omega x - sh^2\omega L}{ch\omega x} = \frac{\theta_0}{ch\omega x} \qquad [6.96]$$

$$\theta(L) = -\theta_0\,(th\omega L\,sh\omega x + ch\omega x) \qquad [6.97]$$

This gives:

$$\theta(L) = T(L) - T_e = \frac{270}{ch\;0.588} = 229.2 \qquad [6.98]$$

$$T(L) = 229.2 + 30 = 259.2C \qquad [6.99]$$

We note that this temperature is much higher than T_0. The flow of losses from the edges is necessarily much lower than would be found if the temperature descended to T_e at the end of the wing.

EXAMPLE 6.3.– Cooling of an IT component

We are intending to cool an IT circuit. The heat can only be evacuated from this circuit through its upper section, whether it is plane or horizontal, which takes the form of a square with sides $a = 5\ cm$ (see Figure 6.3).

The ambient temperature around the circuit is T_e = 25 °C.

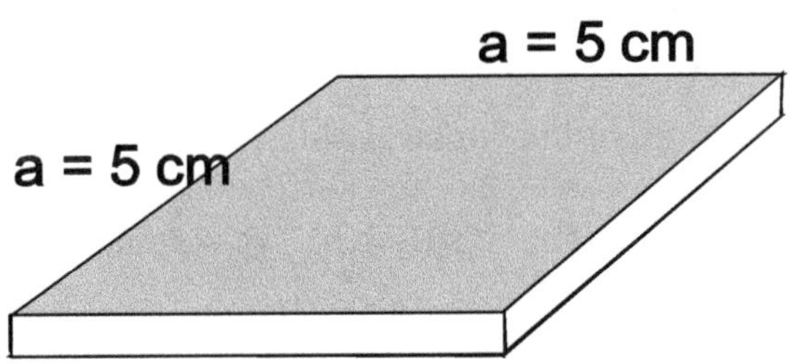

Figure 6.3. *Diagram of the component to be cooled*

1) First, we allow the heat to dissipate by natural convection. The natural convection coefficient is then $h_1 = 8\,W\,m^{-1}\,K^{-1}$.

Knowing that the circuit must not exceed the temperature T_C = 70 °C, what is the maximum flow Φ_1 that can be evacuated by natural convection?

2) Second, a radiator is installed on the circuit, which is made of a square plate A with sides $a = 5\ cm$ and **five copper fins**, vertical and with a height $L = 4\ cm$ and thickness $e = 3\ mm$ (see Figures 6.4 and 6.5).

We will consider that the flow evacuated by each fin is calculated independently of the presence of the other fins (no interactions between the fins). Moreover, the foot of each wing will be at a temperature T_C.

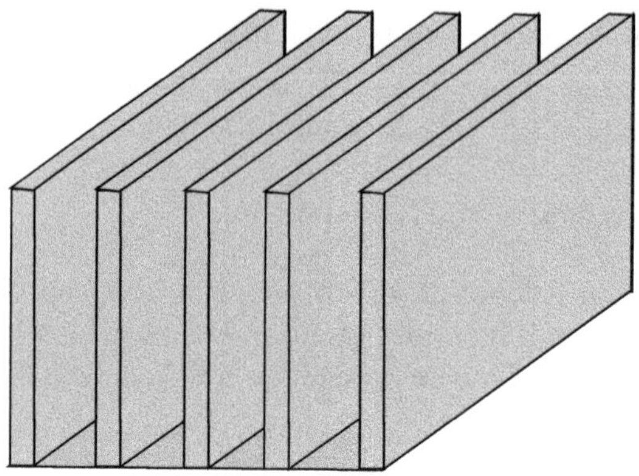

Figure 6.4. *Arrangement of the five radiator fins*

The limit temperature of the circuit is still $T_C = 70\,°C$.

The natural convection coefficient maintains the same value $h_2 = 8\,W\,m^{-1}\,K^{-1}$.

We set the thermal conductibility of copper to be $\lambda_C = 386\ W.m^{-1}.K^{-1}$.

We consider the thermal contact between the circuit and plate A to be perfect.

a) Calculate the flow of heat Φ_A that is evacuated by each fin. To do so, we will neglect the flow of heat emitted by the horizontal surface of each fin.

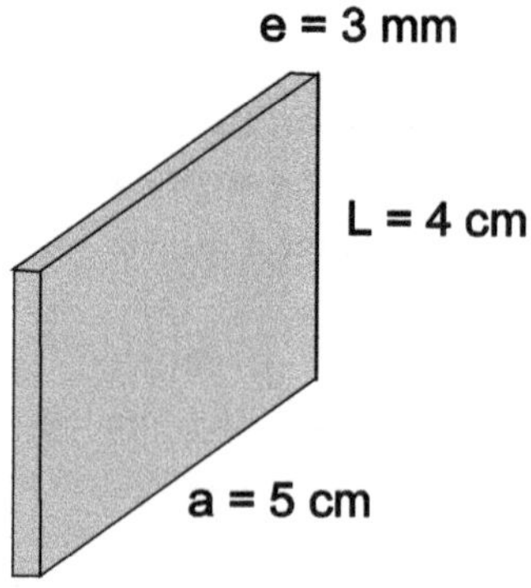

Figure 6.5. *Geometrical characteristics of the fin*

b) Calculate the total flow Φ_2 evacuated by the radiator.

3) We add a ventilator to the system. The convection coefficient then increases to $h_3 = 30\,W\,m^{-1}\,K^{-1}$. Calculate the new value Φ_3 of the total flow that is evacuated by the radiator.

4) The circuit dissipates $P = 19\,W$ by the Joule effect. The limit temperature of the circuit remains T_C = 70 °C.

Among the cooling solutions found in questions 1–3, which are acceptable?

5) We return to the case in question 1, where the flow is evacuated by natural convection. Through a flow $\Phi_4 = 19\,W$ that is dissipated by the circuit, what is the superficial temperature $T_{\max}$ of this circuit? Does this seem acceptable to you?

6) Plate A is, by its very nature, connected to the upper part of the circuit by a thin layer of $x = 0.1\,mm$ of a paste known as "thermal", based on silicon with thermal conductibility $\lambda_P = 0.9\,W.m^{-1}.K^{-1}$. This paste avoids the presence of air inclusions between the element and the plate which, even if it is thin, would be a disturbing thermal resistance. We will denote T_{S2} as the new temperature taken at the foot of each wing.

a) Establish two relations between the flow Φ_5 evacuated and the temperatures T_C and T_{S2}, on the one hand, and the temperatures T_e and T_{S2}, on the other hand, on either side of the layer.

b) Deduce from this the flow evacuated by the radiator in the case where the ventilator is used. Does this value pose a problem?

SOLUTION.–

1) Natural convection only

The exchange surface is:

$$S_C = 25\,cm^2 = 25.10^{-4}\,m^2 \qquad [6.100]$$

The flow evacuated by natural convection will therefore be determined by the maximum temperature and the ambient temperature:

$$\Phi_1 = S h_1 (T_C - T_e) = 25.10^{-4} * 8 * (70 - 25) \quad [6.101]$$

$$\Phi_1 = 0.9\, W \quad [6.102]$$

2) Radiator equipped with fins

a) Evacuation via a single fin

We consider the case of a parallelepiped wing of length L. The flow will be neglected at the end.

The distribution of the temperature along the fin will be given by:

$$\theta = \theta_0\, th(\omega L) \quad [6.103]$$

The flow at the origin of the fin (flow removed from the body to be cooled) will be given by:

$$\Phi = \lambda S \omega \theta_0\, th(\omega L) \quad = \quad \sqrt{h p_e \lambda S}\; \theta_0\, th(\omega L) \quad [6.104]$$

where:

$$\theta_0 = T_C - T_e = 45\,°\text{C} \quad [6.105]$$

We calculate the characteristics of the fin:

Perimeter:

$$p_e = 5 + 5 + 0.3 + 0.3 = 10.6\, cm = 0.106\, m \quad [6.106]$$

Surface area:

$$s = 5 * 0.3 = 1.5\, cm^2 = 1.5.10^{-4}\, m^2 \quad [6.107]$$

Parameter:

$$\omega = \sqrt{\frac{h\,p}{\lambda_C\, s}} = \sqrt{\frac{8*0.106\,p}{386*1.5.10^{-4}}} = 3.83 \quad [6.108]$$

$$\omega L = 3.83*4.10^{-2} = 0.153 \quad [6.109]$$

and

$$th(\omega L) = 0.152 \quad [6.110]$$

$$\sqrt{h\,p_e\,\lambda s} = \sqrt{8*0.106*386*1.5.10^{-4}} = 0.222 \quad [6.111]$$

We deduce from this the flow evacuated by the fin:

$$\Phi_A = \sqrt{h\,p_e\,\lambda S}\;\theta_0\, th(\,\omega L) = 0.222*45*0.152 = 1.518\,W \quad [6.112]$$

b) We will easily deduce from this the total flow Φ_2 evacuated by the radiator.

Overall, the five fins will evacuate:

$$5\,\Phi_A = 1.518*5 = 7.59\,W \quad [6.113]$$

The flow to which the convection of the plane face must be added becomes:

$$\Phi_2 = 7.59 + 0.9 = 8.49\,W \quad [6.114]$$

3) We add a forced convection device

We recalculate the characteristics of the fin:

Perimeter:

$$p_e = 5 + 5 + 0.3 + 0.3 = 10.6\,cm = 0.106\,m \quad [6.115]$$

Surface area:

$$s = 5 * 0.3 = 1.5\ cm^2 = 1.5.10^{-4}\ m^2 \qquad [6.116]$$

Parameter:

$$\omega = \sqrt{\frac{h\,p}{\lambda_C\, s}} = \sqrt{\frac{30 * 0.106}{386 * 1.5.10^{-4}}} = 7.41 \qquad [6.117]$$

$$\omega L = 7.41 * 4.10^{-2} = 0.296 \qquad [6.118]$$

and

$$th(\omega L) = 0.288 \qquad [6.119]$$

$$\sqrt{h\,p_e\,\lambda s} = \sqrt{30 * 0.106 * 386 * 1.5.10^{-4}} = 0.429 \qquad [6.120]$$

We deduce from this the flow evacuated by the fin:

$$\Phi_A = \sqrt{h\,p_e\,\lambda S}\ \theta_0\, th(\omega L) = 0.429 * 45 * 0.288 = 5.56\,W \qquad [6.121]$$

Overall, the five fins will evacuate:

$$5\,\Phi_A = 5.56 * 5 = 27.8\,W \qquad [6.122]$$

The flow to which the convection of the plane face must be added becomes:

$$S\,h_1\left(T_C - T_e\right) = 25.10^{-4} * 30 * (70 - 25) = 3.37\,W \qquad [6.123]$$

$$\Phi_3 = 27.8 + 3.37 = 31.17\,W \qquad [6.124]$$

4) The only acceptable solution is the one that evacuates more than $19\,W$; that is, the one that includes the ventilator (question 3) is acceptable.

5) The flow evacuated by natural convection must now be $\Phi_4 = 19\,W$, which will determine $T_{\max}$ via the relationship:

$$\Phi_4 = 19 = S\,h_1\left(T_C - T_e\right) = 25.10^{-4} * 8 * \left(T_{\max} - 25\right) \qquad [6.125]$$

$$T_{\max} = \frac{19}{25.10^{-4} * 8} + 25 = 975\ °\text{C} \qquad [6.126]$$

This temperature is obviously unacceptable for a semiconductor component.

6) An intermediate layer of thermal paste is applied

a) We establish two relationships between the flow Φ_5 evacuated and the temperatures T_C and T_{S2}, on the one hand, and the temperatures T_e and T_{S2}, on the other hand, on either side of the layer.

Passing through the paste layer, the flow creates a difference in the temperature $T_C - T_{S2}$:

$$\Phi_5 = \lambda\frac{T_C - T_{S2}}{x} = \frac{0.9}{10^{-4}} = 9000\left(T_C - T_{S2}\right) \qquad [6.127]$$

In addition, the relationship between the flow and T_e, and T_{S2} remains unmodified. We consider again the previous results:

$$\Phi_5 = 5\sqrt{h\,p_e\,\lambda S}\,\left(T_e - T_{S2}\right)th(\,\omega L) + S\,h_1\left(T_C - T_e\right) \qquad [6.128]$$

$$\Phi_5 = 0.429 * \left(T_e - T_{S2}\right) * 0.288 + 25.10^{-4} * 30 * \left(T_e - T_{S2}\right)$$

$$= \left[0.618 + 7.5.10^{-2}\right]\left(T_e - T_{S2}\right) \qquad [6.129]$$

$$\Phi_5 = 0.693\left(T_e - T_{S2}\right) \qquad [6.130]$$

b) The flow evacuated by the radiator when the ventilator is used is then re-evaluated as:

$$\Phi_5 = 9000\left(T_C - T_{S2}\right) \qquad [6.131]$$

So:

$$T_C - T_{S2} = \frac{\Phi_5}{9000} \quad [6.132]$$

$$\Phi_5 = 0.693\left(T_e - T_{S2}\right) \quad [6.133]$$

So:

$$T_e - T_{S2} = \frac{\Phi_5}{0.693} \quad [6.134]$$

Subtracting the two temperature differences:

$$\left(T_C - T_{S2}\right) - \left(T_e - T_{S2}\right) = \frac{\Phi_5}{0.693} - \frac{\Phi_5}{9000} \quad [6.135]$$

$$T_C - T_e = \Phi_5\left[\frac{1}{0.693} - \frac{1}{9000}\right] = 1.443\,\Phi_5 \quad [6.136]$$

$$\Phi_5 = \frac{T_C - T}{1.443} = \frac{45}{1.443} = 31.2\,W \quad [6.137]$$

This result is identical (with the calculation rounded, it should be slightly less) to the result of question 3. The thermal paste fulfills its purpose well.

Appendices

Appendix 1

Heat Equation of a Three-dimensional System

In this appendix, we present the heat equation in the general case of a three-dimensional system.

A1.1. Reminder: writing the Fourier law

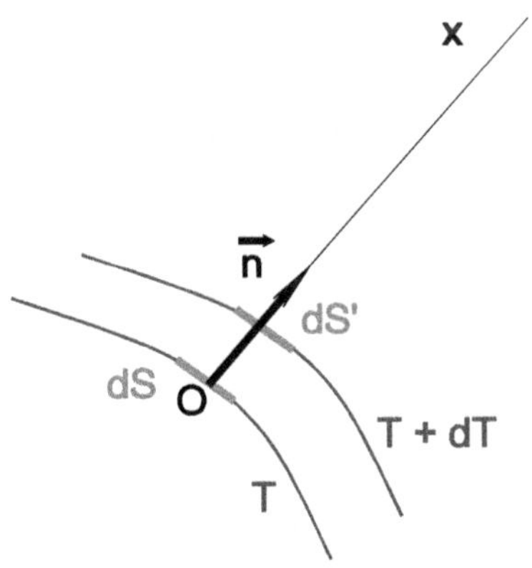

Figure A1.1. *Case of two infinitely close isotherms; relationship with the gradient. For a color version of this figure, see www.iste.co.uk/ledoux/heat1.zip*

Let us consider two isothermal surfaces with respective temperatures T and $T + dT$, which are very close to each other. We have M as a point on the isotherm T. We will consider two infinitely small surfaces dS_1 and dS_2 that are next to each other on each of the isotherms, on M. They have practically the same normal $\vec{n}$, conventionally directed from the surface of temperature T towards the surface of $T + dT$. Locally, we have the situation found in a single-dimensional plane system.

The thermal flow that passes through dS_1 and dS_2 will be expressed by:

$$d\Phi = -\lambda \frac{dT}{dx} dS_1 \qquad [A1.1]$$

where x is a measured axis on the axis bearing $\vec{n}$. We should recall that the minus sign indicates that the heat propagates in the direction of decreasing temperatures. In this expression, the sign implies a positive flow when it flows in the direction of the normal.

Let us consider a surface dS' supported on the same section of cylinder as dS_1 and dS_2, including the normal $\vec{n}'$, make an angle θ with the normal $\vec{n}$ that is common to dS_1 and dS_2.

The same flow of thermal heat obviously passes through dS', dS_1 and dS_2.

We have the known geometric relationship:

$$dS_1 = dS' \cos\theta \qquad [A1.2]$$

and the flow passing through dS' is written as:

$$d\Phi = -\lambda \frac{dT}{dx} dS_1 = = -\lambda \frac{dT}{dx} dS' \cos\theta \qquad [A1.3]$$

Near *M*, the temperature gradient is reduced to a component along the axis of x:

$$\left(grad\vec{\ } T\right)_x = \lambda \frac{dT}{dx} \qquad [A1.4]$$

which can be rewritten as:

$$\left\| \vec{grad}\, T \right\| = \left| \lambda \frac{d\,T}{d\,x} \right| \qquad [A1.5]$$

We note that:

$$\left\| \vec{n}\,' \right\| = 1 \qquad [A1.6]$$

In the end, the flow will take the form of a scalar product:

$$d\,\Phi = -\lambda \frac{d\,T}{d\,x} d\,S' \cos\theta = \left\| \vec{grad}\, T \right\| \left\| \vec{n}\,' \right\| \cos\theta\, d\,S'$$

$$= \left\| \vec{grad}\, T \right\| \left\| \vec{n}\,' \right\| \cos\left(\vec{grad}\, T, \vec{n}\,' \right) d\,S' \qquad [A1.7]$$

Taking into account the orientations of the normal that are specified above:

$$d\,\Phi = -\vec{grad}\, T . \vec{n}\,'\, d\,S' \qquad [A1.8]$$

Therefore, the flow of heat that passes through a closed surface will be written as:

$$\Phi = \iint_S \lambda\, \vec{grad}\, T . \vec{n}\, d\,S \qquad [A1.9]$$

By convention, the normals are always orientated towards the external side of the closed surface.

REMARK.– This convention is imposed in order to correctly write the integral theorems of vectorial geometry: the Stokes equation, Ostrogradsky's equation, etc.

All outgoing flow will thus be counted positively in the integral, and all flows entering will be counted negatively. The integral therefore represents a resulting flow balance, positive if the volume contained in the surface loses heat.

The expressions drawn up here now allow us to write different expressions for the thermal balance in a closed surface. The various forms given to these balances will constitute the various forms of the heat equation, the starting point for all thermal conduction problems.

A1.2. Heat equation

Let us consider a region given by a surface *S*. In the most general case, the thermal conductibility of the material λ is variable in *D*.

The balance of flows through *S* is given by

$$\Phi = \iint_S \lambda \, \vec{grad}\, T . \vec{n} \, dS . \qquad [A1.10]$$

This integral is positive if more heat exits than enters, negative if more enters than exits.

In the former case, the flow will contribute, overall, to "cooling" the region *D*. The sensitive quantity of heat integrated into *D* will diminish over a time dt. In the latter case, the flow will contribute, overall, to heating the region *D*. The sensitive quantity of heat integrated into *D* will increase over a time dt.

$$\Phi = \iint_S \lambda \, \vec{grad}\, T . \vec{n} \, dS \qquad [A1.11]$$

expresses heat transfer per unit of time.

The sensitive quantity of heat integrated into D is written as:

$$Q = \iiint_D \rho c T \, d\omega \qquad [A1.12]$$

where ρ is the (local) density of the material and *c* is its (local) specific heat capacity. The balance equation can therefore be written as:

$$\frac{\partial Q}{\partial t} = -\Phi \qquad [A1.13]$$

or

$$\frac{\partial}{\partial t}\iiint_D \rho c T \, d\omega = -\iint_S \lambda \, grad\vec{d}\, T.\vec{n}\, dS \qquad [A1.14]$$

Integrating the derivation with respect to the time in the integral and transforming the integral of the surface into an integral for the volume, we find

$$\iiint_D \rho c \frac{\partial T}{\partial t} d\omega = -\iint_S \lambda \, grad\vec{d}\, T.\vec{n}\, dS = \iiint_D div \, \lambda \, grad\vec{d}\, T \; \omega \qquad [A1.15]$$

Rewritten as:

$$\iiint_D \rho c \frac{\partial T}{\partial t} - div \, \lambda \, grad\vec{d}\, T \; d\omega = 0 \qquad [A1.16]$$

This integral is written on a region that can be of arbitrary shape. It can therefore only be zero if the integrant is zero everywhere. This gives the equation for the point, general and classic shapes of the heat equation:

$$\rho c \frac{\partial T}{\partial t} = div\left(\lambda \, grad\vec{d}\, T\right) \qquad [A1.17]$$

In the case of a stationary problem, the derivative with respect to time disappears, and we obtain:

$$div\left(\lambda \, grad\vec{d}\, T\right) = 0 \qquad [A1.18]$$

Case of constant physical properties

A significantly important case is found in homogeneous environments, with constant physical properties, both in space and in time:

$$\lambda = Cte, \rho = Cte, c = Cte, \qquad [A1.19]$$

The previous expressions are simplified and give forms in which the heat equation is most generally known:

In the stationary regime:

$$\lambda\, div\left(grad\vec{}\, T \right) = 0 \quad [A1.20]$$

Therefore, in the end:

$$div\, grad\vec{}\, T = 0 \quad [A1.21]$$

or, more often:

$$\Delta T = 0 \quad [A1.22]$$

using the relationship:

$$div\, grad\vec{}\, f = \Delta f \quad [A1.23]$$

In the stationary regime, the temperature is a harmonic function, where Δ is the Laplace operator.

In the non-stationary regime:

$$\frac{\partial T}{\partial t} = \frac{\lambda}{\rho c} div\, grad\vec{}\, T \quad [A1.24]$$

$$\frac{\partial T}{\partial t} = a\, \Delta T \quad [A1.25]$$

Here, we see the fundamental parameter of the properties of the material for non-stationary problems, the thermal diffusivity

$$a = \frac{\lambda}{\rho c} \quad [A1.26]$$

Therefore, generally, in problems with constant properties of the material, the coefficient characterizing the material is **only the thermal conductibility**

for stationary problems and the **thermal diffusivity for non-stationary problems**.

Use of the Laplace operator

$$\Delta T = div\, \vec{grad}\, T \qquad [A1.27]$$

is very practical. Writing equations therefore remains valid for all coordinate systems. It is therefore necessary to use the known form of the Laplace operator for each system, in order to have the heat equations in this coordinate system.

Therefore, we can easily rewrite the heat equation in all coordinate systems, in particular, Cartesian, cylindrical or spherical. Obviously, we will find the expressions established in the case of the simplified systems studied in the main text of this book.

Appendix 2

Heat Equation: Writing in the Main Coordinate Systems

The heat equation results from a **balance** between the heat that, by conduction, enters, exits and accumulates in an **elementary volume**.

A2.1. The elementary volume

The geometry of this elementary volume depends on the coordinate system used.

In an *orthonormalized* system (x, y, z), the volume is a parallelepiped with sides (dx, dy, dz). Its volume is $dx.dy.dz$.

In a **cylindrical** system (r, θ, z), the elementary volume is an annular sector, with length dz, included between two cylinders with radius r and $r + dr$, and on a sector with angle $d\theta$. Its volume is $r.dr.d\theta.dz$.

In a **spherical** system (r, Φ, θ), the elementary volume is determined by two spheres of radius r and $r + dr$, longitudes θ and $\theta + d\theta$, and colatitudes Φ and $\Phi + d\Phi$. Its volume is $r^2 \sin\Phi \, dr \, d\theta \, d\Phi$.

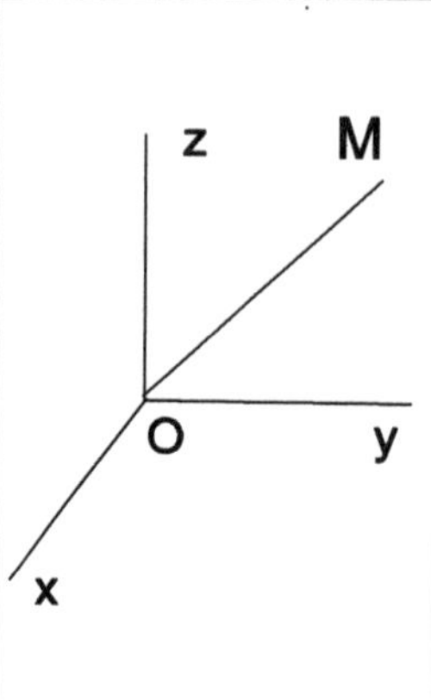	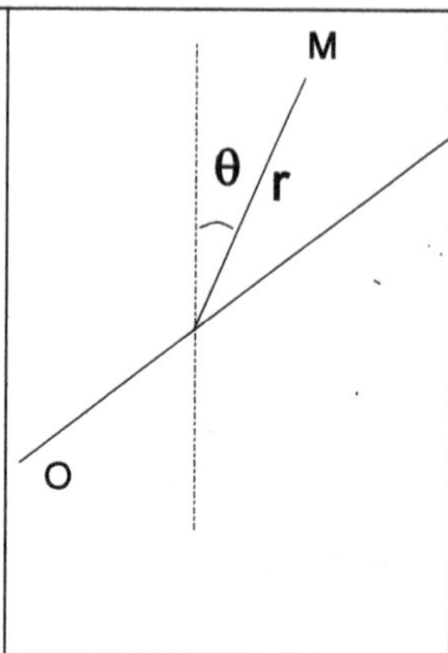	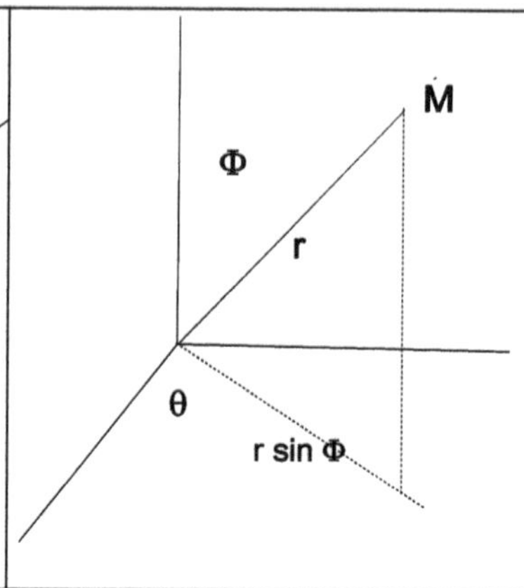
Orthonormalized system	**Cylindrical system**	**Spherical system**
Reference and coordinates	Reference and coordinates	Reference and coordinates
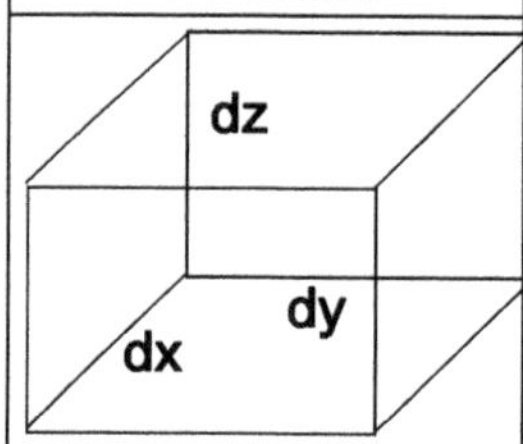	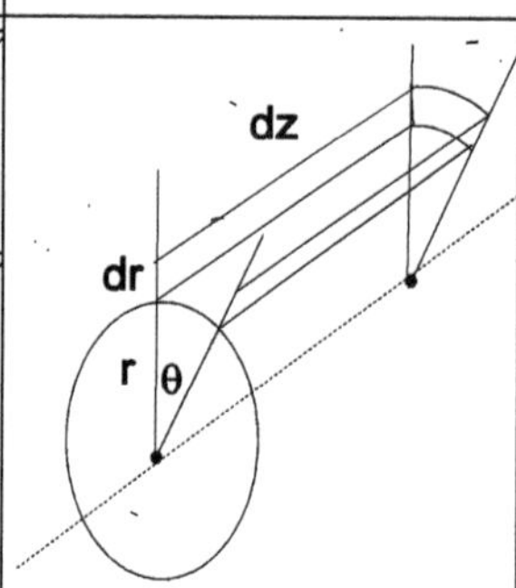	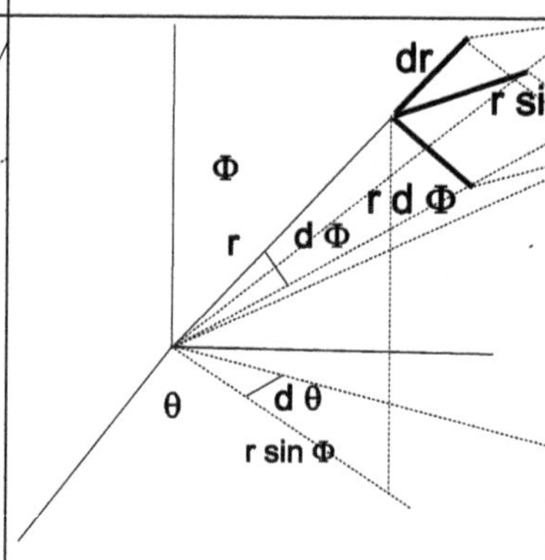
Orthonormalized system	**Cylindrical system**	**Spherical system**
Elementary volume	Elementary volume	Elementary volume

A2.2. *Problem with variable physical properties*

In the most general case, the volumetric mass ρ, the specific heat capacity °C and, in particular, the thermal conductibility λ can vary with temperature. The heat equation is then written as follows:

In the orthonormalized system

$$\rho C\frac{\partial T}{\partial t}=\frac{\partial}{\partial x}\left(\lambda\frac{\partial T}{\partial x}\right)+\frac{\partial}{\partial y}\left(\lambda\frac{\partial T}{\partial y}\right)+\frac{\partial}{\partial z}\left(\lambda\frac{\partial T}{\partial z}\right) \quad [A2.1]$$

In the cylindrical system

$$\rho C\frac{\partial T}{\partial t}=\frac{1}{r}\frac{\partial}{\partial r}\left(\lambda\, r\frac{\partial T}{\partial x}\right)+\frac{1}{r^2}\frac{\partial}{\partial \theta}\left(\lambda\frac{\partial T}{\partial \theta}\right)+\frac{\partial}{\partial z}\left(\lambda\frac{\partial T}{\partial z}\right) \quad [A2.2]$$

In the spherical system

$$\rho C\frac{\partial T}{\partial t} \quad [A2.3]$$

$$=\frac{1}{r^2\sin\Phi}\left[\frac{\partial}{\partial r}\left(\lambda\, r^2\sin\Phi\frac{\partial T}{\partial x}\right)+\frac{1}{\sin\Phi}\frac{\partial}{\partial \theta}\left(\lambda\frac{\partial T}{\partial \theta}\right)+\frac{\partial}{\partial \Phi}\left(\lambda\sin\Phi\,\frac{\partial T}{\partial \Phi}\right)\right] \quad [A2.4]$$

A2.3. Problem with constant physical properties

A2.3.1. *The problem can often be simplified*

If the density ρ, the specific heat capacity °C and, in particular, the thermal conductibility λ are independent of the temperature, we then define the thermal diffusivity a:

$$a=\frac{\lambda}{\rho C} \quad [A2.5]$$

and the heat equation becomes

$$\frac{\partial T}{\partial t}=a\Delta T \quad [A2.6]$$

where ΔT is the Laplace operator for the temperature $T(x,y,z,t)$.

A2.3.2. *This Laplace operator is written, in each of the coordinate systems*

In the orthonormalized system

$$\Delta T = \frac{\partial^2 T}{\partial x^2} + \frac{\partial^2 T}{\partial y^2} + \frac{\partial^2 T}{\partial z^2} \quad [A2.7]$$

In the cylindrical system

$$\Delta T = \frac{1}{r}\frac{\partial}{\partial r} r \frac{\partial T}{\partial x} + \frac{1}{r^2}\frac{\partial^2 T}{\partial \theta^2} + \frac{\partial^2 T}{\partial z^2} \quad [A2.8]$$

In the spherical system

$$\Delta T = \frac{1}{r^2 \sin \Phi}\left[\frac{\partial}{\partial r} r^2 \sin \Phi \frac{\partial T}{\partial x} + \frac{1}{\sin \Phi}\frac{\partial^2 T}{\partial \theta^2} + \frac{\partial}{\partial \Phi} \sin \Phi \frac{\partial T}{\partial \Phi}\right] \quad [A2.9]$$

A2.3.3. *Hence the expressions for the heat equation in a time-dependent regime*

In the orthonormalized system

$$\frac{\partial T}{\partial t} = a\left[\frac{\partial^2 T}{\partial x^2} + \frac{\partial^2 T}{\partial y^2} + \frac{\partial^2 T}{\partial z^2}\right] \quad [A2.10]$$

In the cylindrical system

$$\frac{\partial T}{\partial t} = a\left[\frac{1}{r}\frac{\partial}{\partial r} r \frac{\partial T}{\partial x} + \frac{1}{r^2}\frac{\partial^2 T}{\partial \theta^2} + \frac{\partial^2 T}{\partial z^2}\right] \quad [A2.11]$$

In the spherical system

$$\frac{\partial T}{\partial t} = \frac{a}{r^2 \sin \Phi}\left[\frac{\partial}{\partial r} r^2 \sin \Phi \frac{\partial T}{\partial x} + \frac{1}{\sin \Phi}\frac{\partial^2 T}{\partial \theta^2} + \frac{\partial}{\partial \Phi} \sin \Phi \frac{\partial T}{\partial \Phi}\right] \quad [A2.12]$$

A2.4. Time-independent regime

In numerous cases, the temperature does not depend on the time:

$$\frac{\partial T}{\partial t} = 0 \qquad \text{[A2.13]}$$

A2.4.1. *In the case of variable physical properties*

The heat equation then becomes:

In the orthonormalized system

$$\frac{\partial}{\partial x}\left(\lambda \frac{\partial T}{\partial x}\right) + \frac{\partial}{\partial y}\left(\lambda \frac{\partial T}{\partial y}\right) + \frac{\partial}{\partial y}\left(\lambda \frac{\partial T}{\partial y}\right) = 0 \qquad \text{[A2.14]}$$

In the cylindrical system

$$\frac{1}{r}\frac{\partial}{\partial r}\left(\lambda\, r \frac{\partial T}{\partial x}\right) + \frac{1}{r^2}\frac{\partial}{\partial \theta}\left(\lambda \frac{\partial T}{\partial \theta}\right) + \frac{\partial}{\partial z}\left(\lambda \frac{\partial T}{\partial z}\right) = 0 \qquad \text{[A2.15]}$$

In the spherical system

$$\left[\frac{\partial}{\partial r}\left(\lambda\, r^2 \sin\Phi \frac{\partial T}{\partial x}\right) + \frac{1}{\sin\Phi}\frac{\partial}{\partial \theta}\left(\lambda \frac{\partial T}{\partial \theta}\right) + \frac{\partial}{\partial \Phi}\left(\lambda \sin\Phi \frac{\partial T}{\partial \Phi}\right)\right] = 0 \qquad \text{[A2.16]}$$

A2.4.2. *In the case of constant physical properties*

In this case, the Laplace operator of the temperature is zero. The value of the physical properties no longer intervenes in the general expression for the temperature. Note that the thermal conductivity intervenes in the calculations for flows

$$\Delta T = 0 \qquad \text{[A2.17]}$$

In the orthonormalized system

$$\left[\frac{\partial^2 T}{\partial x^2}+\frac{\partial^2 T}{\partial y^2}+\frac{\partial^2 T}{\partial z^2}\right]=0 \qquad [A2.18]$$

In the cylindrical system

$$\left[\frac{1}{r}\frac{\partial}{\partial r}\, r\frac{\partial T}{\partial x}+\frac{1}{r^2}\frac{\partial^2 T}{\partial \theta^2}+\frac{\partial^2 T}{\partial z^2}\right]=0 \qquad [A2.19]$$

In the spherical system

$$\left[\frac{\partial}{\partial r}\, r^2\sin\Phi\frac{\partial T}{\partial x}+\frac{1}{\sin\Phi}\frac{\partial^2 T}{\partial \theta^2}+\frac{\partial}{\partial \Phi}\sin\Phi\,\frac{\partial T}{\partial \Phi}\right]=0 \qquad [A2.20]$$

A2.5. Writing the Fourier law. How can the expression for the gradient be found?

The Fourier law is written using the temperature gradient:

$$d\varphi=\lambda\, \overrightarrow{grad}\, T.\vec{n}\, dS \qquad [A2.21]$$

It is therefore useful to know how to find the expression for the gradient in various coordinate systems. To do so, we need to know:

a) How to write the expression for the components of the small displacement vector $d\vec{l}$ due to an elementary modification of the coordinates.

This gives a vector of components.

In the orthonormalized system

$$d\vec{l}\left|\begin{matrix} dx \\ dy \\ dz \end{matrix}\right. \qquad [A2.22]$$

In the cylindrical system

$$d\vec{l}\left|\begin{array}{l} dr \\ r\,d\theta \\ dz \end{array}\right. \qquad [A2.23]$$

In the spherical system

$$d\vec{l}\left|\begin{array}{c} dr \\ r\,d\Phi \\ r\sin\Phi\,d\theta \end{array}\right. \qquad [A2.24]$$

b) Recall that the temperature gradient is normal at the isotherms.

If $d\vec{l}$ is a vector that is a tangent to the surface isotherm $T = Cte$, the scalar product of the gradient and of $d\vec{l}$ will be zero.

In the displacement $d\vec{l}$, T therefore does not vary. The exact total differential dT is zero.

Writing in each system of coordinates, we will find $\overrightarrow{grad}\,T$ leading to the components of $d\vec{l}$ in the differential dT.

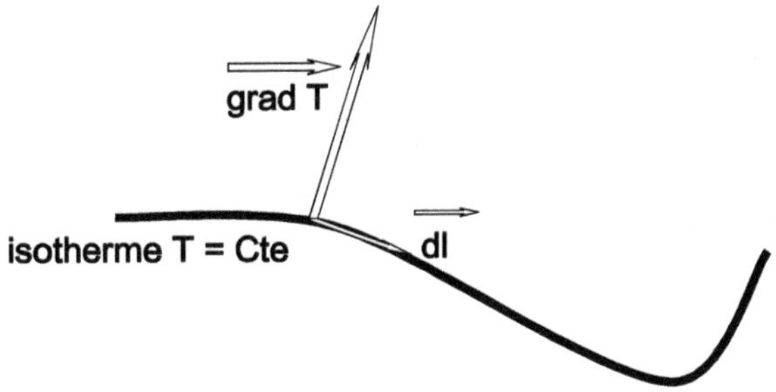

Figure A2.1. *Relationship of the gradient to the isotherm, a particular case of the relationship of the gradient to the equipotential*

In the orthonormalized system

$$dT = \frac{\partial T}{\partial x}dx + \frac{\partial T}{\partial y}dy + \frac{\partial T}{\partial z}dz = \overrightarrow{grad}\,T \,.\, d\vec{l} = 0\,. \qquad [A2.25]$$

Therefore, the components of the gradient are:

$$\overrightarrow{grad}\,T \left| \begin{array}{l} \dfrac{\partial T}{\partial x} \\ \dfrac{\partial T}{\partial y} \\ \dfrac{\partial T}{\partial z} \end{array} \right. \qquad [A2.26]$$

In the cylindrical system

$$dT = \frac{\partial T}{\partial r}dr + \frac{\partial T}{\partial \theta}d\theta + \frac{\partial T}{\partial z}dz \qquad [A2.27]$$

$$= \frac{\partial T}{\partial r}dr + \frac{1}{r}\frac{\partial T}{\partial \theta}r\,d\theta + \frac{\partial T}{\partial z}dz = \overrightarrow{grad}\,T \,.\, d\vec{l} = 0 \qquad [A2.28]$$

Therefore, the components of the gradient in the cylindrical system are:

$$\overrightarrow{grad}\,T \left| \begin{array}{l} \dfrac{\partial T}{\partial r} \\ \dfrac{1}{r}\dfrac{\partial T}{\partial \theta} \\ \dfrac{\partial T}{\partial z} \end{array} \right. \qquad [A2.29]$$

In the spherical system

$$dT = \frac{\partial T}{\partial r}dr + \frac{\partial T}{\partial \Phi}d\Phi + \frac{\partial T}{\partial \theta}d\theta \qquad [A2.30]$$

$$= \frac{\partial T}{\partial r} dr + \frac{1}{r}\frac{\partial T}{\partial \Phi} r\,d\Phi + \frac{1}{r\sin\Phi}\frac{\partial T}{\partial z} r\sin\Phi\,d\theta = \overrightarrow{grad}\,T\,.\,d\vec{l} = 0 \qquad [A2.31]$$

Therefore, the components of the gradient in the spherical system are:

$$\overrightarrow{grad}\,T \quad \begin{matrix} \dfrac{\partial T}{\partial r} \\[2mm] \dfrac{1}{r}\dfrac{\partial T}{\partial \Phi} \\[2mm] \dfrac{1}{r\sin\Phi}\dfrac{\partial T}{\partial \theta} \end{matrix} \qquad [A2.32]$$

Appendix 3

One-dimensional Heat Equation

In this appendix, we present the single-dimensional heat equation in the case of cylindrical and spherical geometries. As specified in section 2.3, we have included the details of deriving this heat equation for these two cases.

A3.1. Case of an axisymmetric system

In this case, we will no longer talk about an Ox axis, but **we direct each radius positively from the axis of the cylinder towards the outside of the cylinder.** Reasoning in the same way as before, we obtain the following conclusions.

Therefore, we have a single-dimensional field. The temperature depends only on *r*, the distance from a point at the axis. **The isotherms are therefore cylinders with radius** r **:** $T = T(r)$.

We no longer reason in m² of surface area, but we reason **for a length of 1 meter of cylinder**[1].

The flow per meter of cylinder through a cylinder of radius r **and of a unit length** is then

$-\lambda \frac{dT}{dr} 2\pi r L$, with $L = 1$, that we will no longer write out in the following.

1 Indeed, in the rest of the calculation, we see that the surface through which the conduction takes place is proportional to *r* and is therefore going to remain within the derivative.

$\frac{dT}{dr} < 0$ when the heat goes from the axis towards the outside, and $-\lambda \frac{dT}{dr}$ is positive when the heat goes in the direction of increasing r.

$\frac{dT}{dx} > 0$ when the heat goes from the outside towards the axis, and $-\lambda \frac{dT}{dr}$ is negative when the heat goes in the direction of decreasing r.

Calculate the change in $-\lambda \frac{dT}{dr} 2\pi r$ when we go from r to $r + dr$, with dr being infinitely small:

$$-\lambda \frac{dT}{dr} 2\pi r \text{ becomes } \lambda \frac{dT}{dr} 2\pi r + \frac{d}{dr}\left(\lambda 2\pi r \frac{dT}{dr} \right) dr. \qquad [A3.1]$$

And, changing the sign,

$$-\lambda \frac{dT}{dr} 2\pi r \text{ becomes } -\lambda \frac{dT}{dr} 2\pi r + \frac{d}{dr}\left(-\lambda 2\pi r \frac{dT}{dr} \right) dr. \qquad [A3.2]$$

We note that an "r" remains below the second derivative of the last expression.

Hence the need to work on a linear meter of cylinder and not a m^2.

We will now consider a cylinder located at $r + dr$**.** The quantity of heat that passes through it per meter of length of cylinder is:

$$-\lambda \frac{dT}{dr} 2\pi r + \frac{d}{dr}\left(-\lambda 2\pi r \frac{dT}{dr} \right) dr \qquad [A3.3a]$$

Taking up the reasoning set out in r:

$-\lambda\frac{dT}{dr}2\pi r+\frac{d}{dr}\left(-\lambda 2\pi r\frac{dT}{dr}\right)dr$ is positive when the heat goes in the direction of increasing r (s moving away from the axis).

If $-\lambda\frac{dT}{dr}2\pi r+\frac{d}{dr}\left(-\lambda 2\pi r\frac{dT}{dr}\right)dr$ is negative, then the heat goes in the direction of the decreasing r (moving closer to the axis).

By a reasoning that is analogous to the plane problem, we write the thermal balance in the small volume $2\pi r\,dr$ in a time dt:

$$dQ=\rho c 2\pi r\,dr\,dT=\left[-\lambda\frac{dT}{dr}2\pi r+\lambda\frac{dT}{dr}2\pi r+\frac{d}{dr}\left(+\lambda 2\pi r\frac{dT}{dr}\right)dr\right]dt \quad [A3.3b]$$

$$\rho c r\,dr\,dT=\left[\frac{d}{dr}\left(+\lambda r\frac{dT}{dr}\right)dr\right]dt \quad [A3.4]$$

$$\rho c\frac{\partial T}{\partial t}=\frac{1}{r}\frac{\partial}{\partial r}\left(\lambda r\frac{\partial T}{\partial r}\right) \quad [A3.5]$$

with the two specific cases:

Constant physical properties:

$$\frac{\partial T}{\partial t}=\frac{a}{r}\frac{\partial}{\partial r}\left(r\frac{\partial T}{\partial r}\right) \quad [A3.6]$$

Stationary conduction:

$$\frac{d}{dr}\left(r\frac{dT}{dr}\right)=0 \quad [A3.7]$$

We note that r is still under the first derivative, which is characteristic of equations with cylindrical symmetry.

A3.2. Case of a spherical system

In this case, we no longer talk about an axis Ox, but **we positively direct each radius of the axis of the cylinder to the outside of the sphere.** Reasoning in the same way as before, we reach the following conclusions.

We have a single-dimensional field. The temperature only depends on r, the distance from a point at the origin of the coordinates, the center of the considered spherical system. **The isotherms are therefore spheres with radius** r **:** $T = T(r)$.

We now reason in terms of flow passing through a sphere.

The balance will be established for a space between two spheres that are infinitely close to radius r and $r + dr$.

We recall that the surface of a sphere of diameter $d = 2r$ is:

$$S = \pi d^2 = 4\pi r^2 \tag{A3.8}$$

and that its volume is:

$$V = \frac{4\pi r^3}{3} = \frac{\pi d^3}{6} \tag{A3.9}$$

A3.2.1. *The flow through a sphere of radius r*

The flow through a sphere with radius r is then $-\lambda \dfrac{dT}{dr} 4\pi r^2$.

If $\dfrac{dT}{dr} < 0$, the heat goes from the origin towards the outside, and $-\lambda \dfrac{dT}{dr}$ is positive when the heat goes in the direction of increasing r.

If $\dfrac{dT}{dx} > 0$, the heat goes from the outside towards the axis, and $-\lambda \dfrac{dT}{dr}$ is negative when the heat goes in the direction of decreasing r.

When we go from r to $r + dr$, dr is infinitely small, $-\lambda \dfrac{dT}{dr} 4\pi r^2$ becomes:

$$-\lambda \frac{dT}{dr} 4\pi r^2 + \frac{d}{dr}\left(-\lambda 4\pi r^2 \frac{dT}{dr} \right) dr \qquad [A3.10]$$

We note that an "r^2" remains under the second derivative of the last expression.

Let us now consider a sphere of radius $r + dr$**.** The quantity of heat that passes through it per meter of length of the cylinder is:

$$-\lambda \frac{dT}{dr} 4\pi r^2 + \frac{d}{dr}\left(-\lambda 4\pi r^2 \frac{dT}{dr} \right) dr \qquad [A3.11]$$

Using the reasoning made for x

$-\lambda \dfrac{dT}{dr} 4\pi r^2 + \dfrac{d}{dr}\left(-\lambda 4\pi r^2 \dfrac{dT}{dr} \right) dr$ is positive when the heat goes in the direction of increasing r (s moving away from the axis).

Therefore, $-\lambda \dfrac{dT}{dr} 4\pi r^2 + \dfrac{d}{dr}\left(-\lambda 4\pi r^2 \dfrac{dT}{dr} \right) dr$ is negative when the heat moves in the direction of decreasing r (moving closer to the axis).

We will set up the **balance** of incoming and outgoing heat **by conduction** in the space between two spheres of radius r and $r + dr$.

Through the "interior" sphere:

$-\lambda \dfrac{dT}{dr} 4\pi r^2$ is positive if the heat goes in the direction of increasing r.

In this case, the heat enters the space and the **term is counted positively in the balance.**

Through the plane $r + dr$, the incoming heat must also be counted positively in the balance.

Now $-\lambda\dfrac{dT}{dr}2\pi r+\dfrac{d}{dr}\left(-\lambda 2\pi r\dfrac{dT}{dr}\right)dr$ is positive when the heat goes in the direction of increasing r, in other words when the heat exits the annular space.

This is therefore **the opposite term, which will intervene in the balance**, in other words $\lambda\dfrac{dT}{dr}4\pi r^2+\dfrac{d}{dr}\left(\lambda 4\pi r^2\dfrac{dT}{dr}\right)dr$.

Hence the resulting term for conduction in the balance:

$$-\lambda\frac{dT}{dr}4\pi r^2+\lambda\frac{dT}{dr}4\pi r^2+\frac{d}{dx}\left(\lambda 4\pi r^2\frac{dT}{dr}\right)dr=\frac{d}{dr}\left(\lambda 4\pi r^2 r\frac{dT}{dr}\right) \quad \text{[A3.12]}$$

By reasoning in the same way as in the previous problem, we write the thermal balance in the small volume $4\pi r^2 dr$ on a time dt:

$$dQ=\rho c 4\pi r^2 dr\, dT=\left[-\lambda\frac{dT}{dr}4\pi r^2+\lambda\frac{dT}{dr}4\pi r^2+\frac{d}{dx}\left(\lambda 4\pi r^2\frac{dT}{dr}\right)dr\right]dt \quad \text{[A3.13]}$$

$$\rho c 4\pi r^2 dr\, dT=\left[\frac{d}{dx}\left(\lambda 4\pi r^2\frac{dT}{dr}\right)dr\right]dt \quad \text{[A3.14]}$$

$$\rho c\frac{\partial T}{\partial t}=\frac{1}{r^2}\frac{\partial}{\partial r}\left(\lambda r^2\frac{\partial T}{\partial r}\right) \quad \text{[A3.15]}$$

with the two specific cases:

Constant physical properties:

$$\frac{\partial T}{\partial t}=\frac{a}{r^2}\frac{\partial}{\partial r}\left(r^2\frac{\partial T}{\partial r}\right) \quad \text{[A3.16]}$$

Stationary conduction:

$$\frac{d}{dr}\left(r^2\frac{dT}{dr}\right)=0 \qquad \text{[A3.17]}$$

We note that r^2 is maintained within the first derivative, which is characteristic of equations with spherical symmetry.

Appendix 4

Conduction of the Heat in a Non-stationary Regime: Solutions to Classic Problems

Hypotheses:

The media are homogeneous.

Their properties are constant.

Specific heat capacity c, density ρ, thermal conductibility λ and consequently, thermal diffusivity $a = \dfrac{\lambda}{\rho C}$ are constant.

The problems are spatially single-dimensional: the temperature only depends on a variable in space $x \; or \; r$ and on the time t.

The definition of the variable θ (re-dimensioned or reduced) varies from one problem to the next.

The solutions use a composite variable ξ and various special functions.

$$\xi = \frac{x}{2\sqrt{at}} = \frac{x}{\sqrt{4at}} \qquad [A4.1]$$

Error function and extensions, tabulated in **Appendix 5**:

$$erf(\xi) = \frac{2}{\sqrt{\pi}} \int_0^{\xi} e^{-u^2} du \quad \text{[A4.2]}$$

$$erfc(\xi) = 1 - \frac{2}{\sqrt{\pi}} \int_0^{\xi} e^{-u^2} du \quad \text{[A4.3]}$$

$$ierfc(\xi) = \frac{1}{\sqrt{\pi}} \exp(-\xi^2) - \xi\, erfc(\xi) \quad \text{[A4.4]}$$

Problem 1: medium with temperature imposed on a wall

Initial temperature of the environment T_i

Imposed temperature T_e at $x = 0$, at time $t = 0$

$$\theta = \frac{T(x,t) - T_e}{T_i - T_e} \quad \text{[A4.5]}$$

$$\theta = erf(\xi) \quad \text{[A4.6]}$$

Problem 2: medium with flow density φ_0 imposed on a wall

Initial temperature of the environment T_i

Constant density of flows φ_0 imposed at $x = 0$, at time $t = 0$

$$\theta = T(x,t) - T_i \quad \text{[A4.7]}$$

$$\theta = \frac{2\varphi_0}{\lambda} \sqrt{at}\, ierfc(\xi) \quad \text{[A4.8]}$$

Problem 3: medium with sinusoidal variation of temperature imposed on a wall

Initial temperature of the environment T_i

Temperature $T(t)$ imposed at $x=0$

$$T(0,t)=T_i\cos(\omega t) \qquad [A4.9]$$

"Stationary regime" solution:

$$\theta=T(x,t)-T_i \qquad [A4.10]$$

$$\theta=T_i\exp\left(-\sqrt{\frac{\omega}{2a}}x\right)\cos\left(\omega t-\sqrt{\frac{\omega}{2a}}x\right) \qquad [A4.11]$$

amplitude:

$$\exp\left(-\sqrt{\frac{\omega}{2a}}x\right) \qquad [A4.12]$$

phase shift:

$$\sqrt{\frac{\omega}{2a}}x \qquad [A4.13]$$

Problem 4: abrupt contact between two media with respective homogeneous temperature T_1 and T_2

Border between the two environments at $x=0$; contact at time $t=0$

$$\theta_1=\frac{T_1(x,t)-T_{i1}}{T_{i2}-T_{i1}} \qquad [A4.14]$$

$$\theta_2 = \frac{T_2(x,t) - T_{i2}}{T_{i2} - T_{i1}} \qquad [A4.15]$$

Effusivity of media 1 *and* 2

$$E_1 = \sqrt{\lambda_1 \rho_1 C_1} \qquad [A4.16]$$

$$E_2 = \sqrt{\lambda_2 \rho_2 C_2} \qquad [A4.17]$$

medium 1 for $x<0$

$$\theta_1 = \frac{E_1}{E_1 + E_2} erfc(|\xi|) \qquad [A4.18]$$

medium 2 for $x>0$

$$\theta_1 = \frac{E_2}{E_1 + E_2} erfc(\xi)$$

Appendix 5

Table of $erf(x)$, $erfc(x)$ and $ierfc(x)$ Functions

These three functions $erf(x)$, $erfc(x)$ and $ierfc(x)$, are fundamental in processing single-dimensional non-stationary conduction problems.

They are defined by:

$erf(x)$, $erfc(x)$ and $ierfc(x)$ functions

$$erf(x) = \frac{2}{\pi}\int_0^x \exp(-\xi^2)\,d\xi \quad [A5.1]$$

$$erfc(x) = 1 - \frac{2}{\pi}\int_0^x \exp(-\xi^2)\,d\xi \quad [A5.2]$$

$$ierfc(x) = \frac{1}{\sqrt{\pi}}\exp(-x^2) - x\,erfc(x) \quad [A5.3]$$

x	$erf(x)$	$erfc(x)$	$ierfc(x)$		x	$erf(x)$	$erfc(x)$	$ierfc(x)$
0	0	1	0.56418958		**1**	0.84270079	0.15729921	0.05025454
0.05	0.05637198	0.94362802	0.51559947		**1.1**	0.88020507	0.11979493	0.03646538
0.1	0.11246292	0.88753708	0.46982209		**1.2**	0.91031398	0.08968602	0.02604895
0.15	0.16799597	0.83200403	0.42683646		**1.3**	0.93400794	0.06599206	0.01831432

0.2	0.22270259	0.77729741	0.38660791		**1.4**	0.95228512	0.04771488	0.01267002
0.25	0.27632639	0.72367361	0.34908866		**1.5**	0.96610515	0.03389485	0.00862286
0.3	0.32862676	0.67137324	0.31421848		**1.6**	0.97634838	0.02365162	0.00577194
0.35	0.37938205	0.62061795	0.28192557		**1.7**	0.98379046	0.01620954	0.0037993
0.4	0.42839236	0.57160764	0.25212759		**1.8**	0.9890905	0.0109095	0.00245876
0.45	0.47548172	0.52451828	0.22473278		**1.9**	0.99279043	0.00720957	0.00156419
0.5	0.52049988	0.47950012	0.19964123		**2**	0.99532227	0.00467773	0.00097802
0.55	0.56332337	0.43667663	0.17674618		**2.1**	0.99702053	0.00297947	0.00060095
0.6	0.60385609	0.39614391	0.15593537		**2.2**	0.99813715	0.00186285	0.00036282
0.66	0.64937669	0.35062331	0.13354955		**2.3**	0.99885682	0.00114318	0.0002152
0.7	0.67780119	0.32219881	0.12009827		**2.4**	0.99931149	0.00068851	0.00012539
0.75	0.71115563	0.28884437	0.10483226		**2.5**	0.99959305	0.00040695	7.1762E-05
0.8	0.74210096	0.25789904	0.09117366		**2.6**	0.99976397	0.00023603	4.0336E-05
0.85	0.77066806	0.22933194	0.07900271		**2.7**	0.99986567	0.00013433	2.2264E-05
0.9	0.79690821	0.20309179	0.06820168		**2.8**	0.99992499	7.5013E-05	1.2067E-05
0.95	0.82089081	0.17910919	0.05865589		**2.9**	0.9999589	4.1098E-05	6.4216E-06
1	0.84270079	0.15729921	0.05025454		**3**	0.99997791	2.209E-05	3.355E-06

These functions are represented in Figure A5.1.

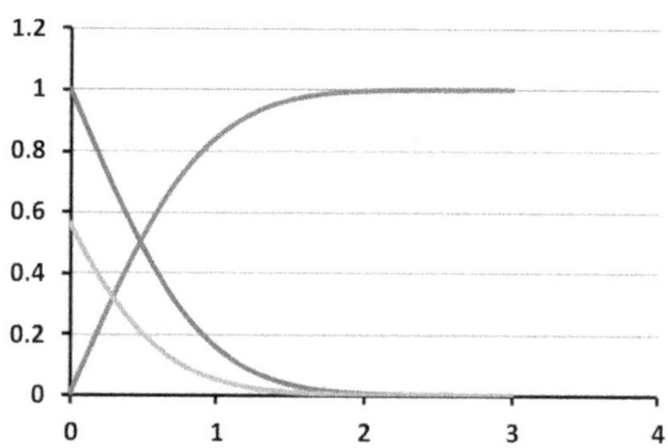

Figure A5.1. *$erf\,\xi$, $erf\,\xi$ and $ierfc\,\xi$ functions. For a color version of this figure, see www.iste.co.uk/ledoux/heat1.zip*

Appendix 6

Complementary Information Regarding Fins

Here, we provide the solutions that result from a single-dimensional conduction model

$$T = T(x)$$

A6.1. Rectangular wings. Solutions to classic problems

Fin with cross-section $e*l$ and length L

S: surface of the cross-section of a fin; $S = e*l$

p_e : perimeter of fin; $p = 2(l+e)$

λ: thermal conductibility of the material that makes up the fin

h: convection coefficient between the fin and the surrounding atmosphere

T_0 : temperature of the fin at its origin (temperature of the body to cool down)

T_e : ambient temperature

We define a parameter ω:

$$\omega=\sqrt{\frac{h\,p_e}{\lambda S}} \tag{A6.1}$$

Relative temperatures:

$$\theta=T(x)-T_e \tag{A6.2a}$$

$$\theta_0=T_0-T_e \tag{A6.2b}$$

Case 1: Infinitely long fin

Distribution of the temperature along the fin:

$$\theta=\theta_0\exp(-\omega x) \tag{A6.3}$$

Flow at the origin of the fin (flow taken from the body to be cooled):

$$\Phi=\lambda S\,\omega\theta_0=\sqrt{h\,p_e\,\lambda S}\;\theta_0 \tag{A6.4}$$

Case 2: Fin of length *L*. Zero flow at the extremity

Distribution of the temperature along the fin:

$$\theta=\theta_0\,th(\,\omega L) \tag{A6.5}$$

Flow at the origin of the fin (flow taken from the body to be cooled):

$$\Phi=\lambda S\,\omega\theta_0\,th(\,\omega L)=\sqrt{h\,p_e\,\lambda S}\;\theta_0\,th(\,\omega L) \tag{A6.6}$$

Case 3: Wing of length *L*. Convection coefficient *h* at the end

Distribution of the temperature along the wing:

$$\theta = \theta_0 \frac{\cosh\left[\omega(L-x)\right] + \frac{h}{\omega\lambda}\sinh\left[\omega(L-x)\right]}{\cosh\left[\omega L\right] + \frac{h}{\omega\lambda}\sinh\left[\omega L\right]} \quad \text{[A6.7]}$$

Flow at the origin of the fin (flow taken from the body to be cooled):

$$\Phi = \lambda S \omega\, \theta_0 \frac{th\left[\omega L\right] + \frac{h}{\omega\lambda}}{1 + \frac{h}{\omega\lambda} th\left[\omega L\right]} = \sqrt{h\, p_e\, \lambda S}\; \theta_0 \frac{th\left[\omega L\right] + \frac{h}{\omega\lambda}}{1 + \frac{h}{\omega\lambda} th\left[\omega L\right]} \quad \text{[A6.8]}$$

Appendix 7

The Laplace Transform

The Laplace transform is a precious tool for the resolution of differential equations with partial derivatives, in particular, with two variables. Indeed, it allows this equation to be transformed into an algebraic equation, where only derivatives with one variable continue to be present.

In problems that depend on time, the Laplace transform is often used with respect to this variable.

It was therefore widely used to establish the main solutions for non-stationary conduction, like in this book, or even in signal theory.

We give little basic information about this method. We have chosen to shorten this for our readers. For most complex problems, numerical methods are applied more often.

The inverse transform, or Mellin–Fourier transform, is not used in practice due to the mathematical difficulties posed by the inversion of the expressions found in physics.

In practice, we compare the result found with the Laplace transform with a table of transforms that were previously established. We find many in the literature.

A7.1. Definition

The Laplace transform transforms a function of a variable, for example, $f(t)$ in another function $F(p)$. We note that F is a function of another variable p.

The function $f(t)$ must be real, locally integrable, defined for $t \geq 0$. It must be limited:

$$|f(t)| < K e^{at} \qquad [A7.1a]$$

It is defined by:

$$f(t) \to F(p) = \int_0^{\infty} e^{-pt} f(t) dt \qquad [A7.1b]$$

The notation

$$\to F(p) = £\ f(t), \qquad [A7.2]$$

or an analogue, is often adopted. We often write the origin function with a lowercase letter and the destination function in capital letters, whether the alphabet used is Roman or Greek. This is not an absolute law, but a practical convention that we will adopt here.

The properties of the transforms of derivatives and primitives of the origin function make the transform particularly useful.

A7.2. Derivatives and integrals

Transform of a derivative with respect to time $\dfrac{df(t)}{dt}$:

$$£\frac{df(t)}{dt} = \int_0^{\infty} e^{-pt} \frac{df(t)}{dt} dt = e^{-pt} f(t)\Big|_0^{\infty} - \int_0^{\infty} -pe^{-pt} f(t)\, dt \qquad [A7.3]$$

$$£\frac{df(t)}{dt} = -f(0) + p\int_0^{\infty} e^{-pt} f(t)\, dt \qquad [A7.4]$$

after an integration by parts.

We therefore have an important property:

$$£\frac{d f(t)}{dt} = -f(0) + pF(p) \qquad [A7.5]$$

We see the presence of the term $-f(0)$. In many cases, $f(0)=0$, but we must always keep the existence of this term in mind in each new problem, in the case where this condition of nullity at the origin of the signal time is not fulfilled. We also see why it is important that the function is limited, beyond the fact that this property guarantees the convergence of the definition integral.

Transform of an integral in time $\int f(t)dt$: [A7.6]

$$£\int f(t)dt = \int_0^\infty e^{-pt} \int f(t)dt\, dt = \frac{F(p)}{p} \qquad [A7.7]$$

A7.3. We will give two examples

A7.3.1. *Transform of the Heaviside step function*

The Heaviside function is defined by:

$$t \leq 0 \quad ; \quad H(t) = 0 \qquad [A7.8]$$

$$t > 0 \quad ; \quad H(t) = 1 \qquad [A7.9]$$

Its transform will be:

$$H(t) \rightarrow F(p) = \int_0^\infty e^{-pt} dt = \left. \frac{-e^{-pt}}{p} \right|_0^\infty = \frac{1}{p} \qquad [A7.10]$$

In the same way, all multiples of the Heaviside function, such that

$$t \leq 0 \quad ; \quad f(t) = 0 \qquad [A7.11]$$

$$t > 0 \quad ; \quad f(t) = a \qquad [A7.12]$$

will be transformed into:

$$f(t) \rightarrow F(p) = \int_0^\infty a e^{-pt} \, dt = \left. \frac{-ae^{-pt}}{p} \right|_0^\infty = \frac{a}{p} \qquad [A7.13]$$

A7.3.2. *Transform of the erfc function*

We will simply give the result with a view to a later calculation

The function $f(t,x) = erfc \dfrac{\alpha x}{2\sqrt{t}}$

is transformed into $F(p) = \dfrac{\exp - \alpha x \sqrt{p}}{p}$ [A7.14]

A7.4. Resolution of a problem of single-dimensional non-stationary transfer

The interest of the transform appears for a differential equation with partial derivatives. For example, for the heat equation:

$$\frac{\partial f(t,x)}{\partial t} = a \frac{\partial^2 f(t,x)}{\partial x^2} \quad \rightarrow \quad pF(p,x) = a \frac{\partial^2 F(p,x)}{\partial x^2} \qquad [A7.15]$$

We see that $F(p,x)$, the transform for $f(t,x)$, responds to a differential equation with one variable.

It will then be necessary to also transform the limit conditions.

We will apply this method to resolving a problem to which we have already found the solution in section 5.1.1, in other words the non-stationary conductive transfer at a fixed wall temperature at the initial instant.

This time we are looking for the solution via a new reduced temperature:

$$\theta(x,t) = \frac{T - T_i}{T_e - T_i} \quad [A7.16]$$

The notations for the temperatures remain as those in section 5.1.1.

This new definition is motivated by the desire to have an unknown that is canceled out at the time origin.

Answers the equation

$$\frac{\partial f(x,t)}{\partial t} = a\frac{\partial^2 f(x,t)}{\partial x^2} \quad [A7.17]$$

with the conditions to fulfill:

$$\theta(x,0) = 0 \quad [A7.18]$$

$$\theta(0,t) = 1 \quad [A7.19]$$

$$\theta(\infty,t) = 0 \quad [A7.20]$$

$\theta(x,t)$ is transformed into $\Theta(x,p)$

$$\pounds\theta(x,t) = \Theta(x) \quad [A7.21]$$

$$\theta(\infty,t) = 0 \quad [A7.22]$$

We apply the Fourier transform to the equation:

$$\frac{\partial^2 \Theta(x)}{\partial x^2} - \frac{p}{a}\Theta(x) = 0 \quad [A7.23]$$

taking into account:

$$\pounds\frac{\partial \theta}{\partial t} = p\,\Theta - \theta(t=0) = p\,\Theta \quad [A7.24]$$

The integration provides a general solution:

$$\Theta(x,p) = A\exp\sqrt{\frac{p}{a}}x + B\exp-\sqrt{\frac{p}{a}}x \qquad [A7.25]$$

The conditions are transformed into:

$\theta(0,t)=1$ gives:

$$\pounds\theta(0,t) = \Theta(0,t) = \frac{1}{p} \qquad [A7.26]$$

At infinity, $\Theta(x,p)$ must remain finite.

The constants are determined by these conditions.

At infinity, the positive exponential is divergent; therefore: $A=0$

$\Theta(0,p)=\dfrac{1}{p}$, therefore:

$$\Theta(0,p) = B\exp-\sqrt{\frac{p}{a}}0 = \frac{1}{p} \qquad [A7.27]$$

and:

$$B = \frac{1}{p} \qquad [A7.28]$$

Therefore, the following remains:

$$\Theta(x,p) = \frac{\exp-\sqrt{\frac{p}{a}}x}{p} \qquad [A7.29]$$

which is of the form $F(p)=\dfrac{\exp-\alpha x\sqrt{p}}{p}$ [A7.30] with $\alpha = \dfrac{1}{\sqrt{a}}$ [A7.31]

The solution to the problem will therefore be:

$$\theta(x,t) = erfc\frac{x}{\sqrt{4at}} \quad [A7.32]$$

where we recognize the variable $\xi(x,t) = \dfrac{x}{\sqrt{4at}}$ [A7.33]

which is very coherent with the solution found in section 5.1.1:

$$\frac{T - T_e}{T_i - T_e} = -\frac{T - T_i + T_i - T_e}{T_e - T_i} = 1 - \frac{T - T_i}{T_e - T_i} = 1 - erfc\,\xi = erf\,\xi \quad [A7.34]$$

Appendix 8

Reminders Regarding Hyperbolic Functions

Hyperbolic functions are defined by:

hyperbolic sinus $shx = \dfrac{e^x - e^{-x}}{2}$ [A8.1]

hyperbolic cosinus $chx = \dfrac{e^x + e^{-x}}{2}$ [A8.2]

hyperbolic tangent $thx = \dfrac{shx}{chx} = \dfrac{e^x - e^{-x}}{e^x + e^{-x}}$ [A8.3]

We can also define a hyperbolic cotangent

$$\coth x = \frac{1}{thx} = \frac{chx}{shx} = \frac{e^x + e^{-x}}{e^x - e^{-x}} \qquad [A8.4]$$

The denomination of hyperbolic sinus, cosinus and tangent comes from the relative similarity of the relationships found between these functions, with the same relationships found for sinusoidal functions.

It must be noted that the similarity stops at these relationships and that the form of the hyperbolic functions is totally different from that of the sinusoidal functions.

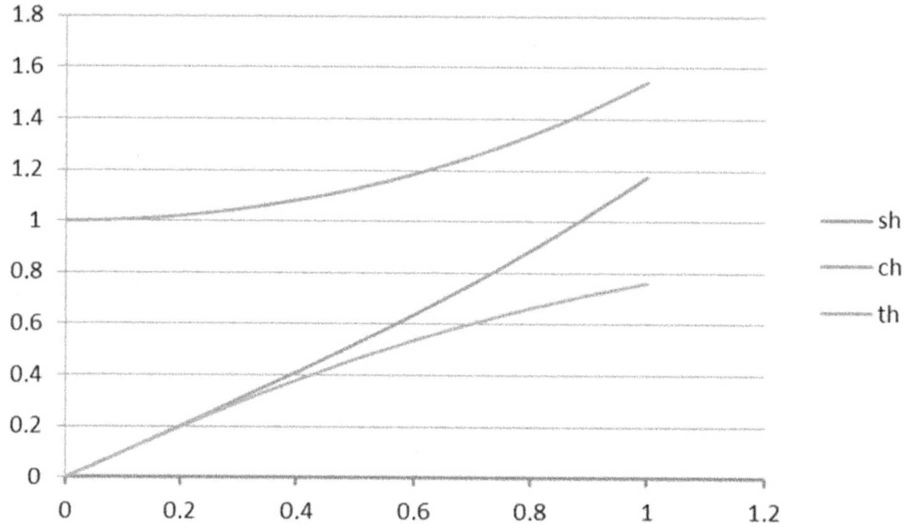

Figure A8.1. *Representation of the hyperbolic functions. For a color version of this figure, see www.iste.co.uk/ledoux/heat1.zip*

The main relationships to know are:

$$ch^2 x - sh^2 x = 1 \quad [A8.5]$$

$$sh(-x) = -sh\,x \quad [A8.6]$$

$$ch(-x) = ch\,x \quad [A8.7]$$

$$th(-x) = -th\,x \quad [A8.8]$$

$$\frac{d\,sh\,x}{d\,x} = ch\,s \quad [A8.9]$$

$$\frac{d\,ch\,x}{d\,x} = sh\,x \quad [A8.10]$$

$$\frac{d\,th\,x}{d\,x} = \frac{1}{ch^2 x} \quad [A8.11]$$

$$sh(x+y) = sh\,x\,ch\,y + ch\,x\,sh\,y \qquad [A8.12]$$

$$ch(x+y) = ch\,x\,ch\,y + sh\,x\,sh\,y \qquad [A8.13]$$

which implies:

$$sh(x-y) = sh\,x\,ch\,y - ch\,x\,sh\,y \qquad [A8.14]$$

$$ch(x-y) = ch\,x\,ch\,y - sh\,x\,sh\,y \qquad [A8.15]$$

References

[BYR 66] BYRON BIRD R., STEWART W.E., LIGHTFOOT, E.N., *Transport Phenomena*, John Wiley & Sons, New York, 1966.

[COS 97] COSAR P., *Aide mémoire du thermicien*, Amicale des anciens élèves de l'Ecole de thermique (ed.), Elsevier, Paris, 1997.

[GOS 93] GOSSE J., *Guide technique de thermique*, Dunod, Paris, 1993.

[HOL 86] HOLMAN J.P., *Heat Transfer*, McGraw Hill, New York, 1986.

[JAN 16] JANNOT Y., MOYNE C., *Transferts thermiques*, Edilivres, Paris, 2016.

[KNU 58] KNUDSEN J.G., KATZ D.L., *Fluid Dynamics and Heat Transfer*, McGraw Hill, New York, 1958.

[LEO 79] LEONTIEV A., *Théorie des échanges de chaleur et de masse*, Editions MIR, Moscow, 1979.

[PER 97] PEREZ J.P., *Thermodynamique, fondements et applications*, Masson, Paris, 1997.

[ROH 98] ROHSENOW, W.M., HARTNETT, J.P., CHO, Y.I., *Handbook of Heat Transfer*, McGraw Hill, New York, 1998.

[SAC 15] SACCADURA J.F., *Transferts thermiques, initiation et approfondissement*, Lavoisier Editeur, Paris, 2015.

[VAN 76] VAN WYLEN G., SONNTAG R.E., *Fundamentals of Classical Thermodynamics*, SI Vers 2, John Wiley & Sons, New York, 1976.

Index

I, R, T

Other titles from

iSTE

in

Mechanical Engineering and Solid Mechanics

2021

DAHOO Pierre Richard, POUGNET Philippe, EL HAMI Abdelkhalak
Applications and Metrology at Nanometer Scale 1: Smart Materials, Electromagnetic Waves and Uncertainties
(Reliability of Multiphysical Systems Set – Volume 9)

2020

SALENÇON Jean
Elastoplastic Modeling

2019

BAYLE Franck
Reliability of Maintained Systems Subjected to Wear Failure Mechanisms: Theory and Applications
(Reliability of Multiphysical Systems Set – Volume 8)

BEN KAHLA Rabeb, BARKAOUI Abdelwahed, MERZOUKI Tarek
Finite Element Method and Medical Imaging Techniques in Bone Biomechanics
(Mathematical and Mechanical Engineering Set – Volume 8)

IONESCU Ioan R., QUEYREAU Sylvain, PICU Catalin R., SALMAN Oguz Umut
Mechanics and Physics of Solids at Micro- and Nano-Scales

LE VAN Anh, BOUZIDI Rabah
Lagrangian Mechanics: An Advanced Analytical Approach

MICHELITSCH Thomas, PÉREZ RIASCOS Alejandro, COLLET Bernard, NOWAKOWSKI Andrzej, NICOLLEAU Franck
Fractional Dynamics on Networks and Lattices

SALENÇON Jean
Viscoelastic Modeling for Structural Analysis

VÉNIZÉLOS Georges, EL HAMI Abdelkhalak
Movement Equations 5: Dynamics of a Set of Solids
(Non-deformable Solid Mechanics Set – Volume 5)

2018

BOREL Michel, VÉNIZÉLOS Georges
Movement Equations 4: Equilibriums and Small Movements
(Non-deformable Solid Mechanics Set – Volume 4)

FROSSARD Etienne
Granular Geomaterials Dissipative Mechanics: Theory and Applications in Civil Engineering

RADI Bouchaib, EL HAMI Abdelkhalak
Advanced Numerical Methods with Matlab® 1: Function Approximation and System Resolution
(Mathematical and Mechanical Engineering SET – Volume 6)
Advanced Numerical Methods with Matlab® 2: Resolution of Nonlinear, Differential and Partial Differential Equations
(Mathematical and Mechanical Engineering SET – Volume 7)

SALENÇON Jean
Virtual Work Approach to Mechanical Modeling

2017

BOREL Michel, VÉNIZÉLOS Georges
Movement Equations 2: Mathematical and Methodological Supplements
(Non-deformable Solid Mechanics Set – Volume 2)
Movement Equations 3: Dynamics and Fundamental Principle
(Non-deformable Solid Mechanics Set – Volume 3)

BOUVET Christophe
Mechanics of Aeronautical Solids, Materials and Structures
Mechanics of Aeronautical Composite Materials

BRANCHERIE Delphine, FEISSEL Pierre, BOUVIER Salima,
IBRAHIMBEGOVIC Adnan
From Microstructure Investigations to Multiscale Modeling:
Bridging the Gap

CHEBEL-MORELLO Brigitte, NICOD Jean-Marc, VARNIER Christophe
From Prognostics and Health Systems Management to Predictive
Maintenance 2: Knowledge, Traceability and Decision
(Reliability of Multiphysical Systems Set – Volume 7)

EL HAMI Abdelkhalak, RADI Bouchaib
Dynamics of Large Structures and Inverse Problems
(Mathematical and Mechanical Engineering Set – Volume 5)
Fluid-Structure Interactions and Uncertainties: Ansys and Fluent Tools
(Reliability of Multiphysical Systems Set – Volume 6)

KHARMANDA Ghias, EL HAMI Abdelkhalak
Biomechanics: Optimization, Uncertainties and Reliability
(Reliability of Multiphysical Systems Set – Volume 5)

LEDOUX Michel, EL HAMI Abdelkhalak
Compressible Flow Propulsion and Digital Approaches in Fluid Mechanics
(Mathematical and Mechanical Engineering Set – Volume 4)
Fluid Mechanics: Analytical Methods
(Mathematical and Mechanical Engineering Set – Volume 3)

MORI Yvon
Mechanical Vibrations: Applications to Equipment

2016

BOREL Michel, VÉNIZÉLOS Georges
Movement Equations 1: Location, Kinematics and Kinetics
(Non-deformable Solid Mechanics Set – Volume 1)

BOYARD Nicolas
Heat Transfer in Polymer Composite Materials

CARDON Alain, ITMI Mhamed
New Autonomous Systems
(Reliability of Multiphysical Systems Set – Volume 1)

DAHOO Pierre Richard, POUGNET Philippe, EL HAMI Abdelkhalak
Nanometer-scale Defect Detection Using Polarized Light
(Reliability of Multiphysical Systems Set – Volume 2)

DE SAXCÉ Géry, VALLÉE Claude
Galilean Mechanics and Thermodynamics of Continua

DORMIEUX Luc, KONDO Djimédo
Micromechanics of Fracture and Damage
(Micromechanics Set – Volume 1)

EL HAMI Abdelkhalak, RADI Bouchaib
Stochastic Dynamics of Structures
(Mathematical and Mechanical Engineering Set – Volume 2)

GOURIVEAU Rafael, MEDJAHER Kamal, ZERHOUNI Noureddine
From Prognostics and Health Systems Management to Predictive Maintenance 1: Monitoring and Prognostics
(Reliability of Multiphysical Systems Set – Volume 4)

KHARMANDA Ghias, EL HAMI Abdelkhalak
Reliability in Biomechanics
(Reliability of Multiphysical Systems Set –Volume 3)

MOLIMARD Jérôme
Experimental Mechanics of Solids and Structures

RADI Bouchaib, EL HAMI Abdelkhalak
Material Forming Processes: Simulation, Drawing, Hydroforming and Additive Manufacturing
(Mathematical and Mechanical Engineering Set – Volume 1)

2015

KARLIČIĆ Danilo, MURMU Tony, ADHIKARI Sondipon, MCCARTHY Michael
Non-local Structural Mechanics

SAB Karam, LEBÉE Arthur
Homogenization of Heterogeneous Thin and Thick Plates

2014

ATANACKOVIC M. Teodor, PILIPOVIC Stevan, STANKOVIC Bogoljub, ZORICA Dusan
Fractional Calculus with Applications in Mechanics: Vibrations and Diffusion Processes
Fractional Calculus with Applications in Mechanics: Wave Propagation, Impact and Variational Principles

CIBLAC Thierry, MOREL Jean-Claude
Sustainable Masonry: Stability and Behavior of Structures

ILANKO Sinniah, MONTERRUBIO Luis E., MOCHIDA Yusuke
The Rayleigh−Ritz Method for Structural Analysis

LALANNE Christian
Mechanical Vibration and Shock Analysis – 5-volume series – 3rd edition
Sinusoidal Vibration – Volume 1
Mechanical Shock – Volume 2
Random Vibration – Volume 3
Fatigue Damage – Volume 4
Specification Development – Volume 5

LEMAIRE Maurice
Uncertainty and Mechanics

2013

ADHIKARI Sondipon
Structural Dynamic Analysis with Generalized Damping Models: Analysis

ADHIKARI Sondipon
Structural Dynamic Analysis with Generalized Damping Models: Identification

BAILLY Patrice
Materials and Structures under Shock and Impact

BASTIEN Jérôme, BERNARDIN Frédéric, LAMARQUE Claude-Henri
Non-smooth Deterministic or Stochastic Discrete Dynamical Systems: Applications to Models with Friction or Impact

EL HAMI Abdelkhalak, RADI Bouchaib
Uncertainty and Optimization in Structural Mechanics

KIRILLOV Oleg N., PELINOVSKY Dmitry E.
Nonlinear Physical Systems: Spectral Analysis, Stability and Bifurcations

LUONGO Angelo, ZULLI Daniele
Mathematical Models of Beams and Cables

SALENÇON Jean
Yield Design

2012

DAVIM J. Paulo
Mechanical Engineering Education

DUPEUX Michel, BRACCINI Muriel
Mechanics of Solid Interfaces

ELISHAKOFF Isaac *et al.*
Carbon Nanotubes and Nanosensors: Vibration, Buckling and Ballistic Impact

GRÉDIAC Michel, HILD François
Full-Field Measurements and Identification in Solid Mechanics

GROUS Ammar
Fracture Mechanics – 3-volume series
Analysis of Reliability and Quality Control – Volume 1
Applied Reliability – Volume 2
Applied Quality Control – Volume 3

RECHO Naman
Fracture Mechanics and Crack Growth

2011

KRYSINSKI Tomasz, MALBURET François
Mechanical Instability

SOUSTELLE Michel
An Introduction to Chemical Kinetics

2010

BREITKOPF Piotr, FILOMENO COELHO Rajan
Multidisciplinary Design Optimization in Computational Mechanics

DAVIM J. Paulo
Biotribolgy

PAULTRE Patrick
Dynamics of Structures

SOUSTELLE Michel
Handbook of Heterogenous Kinetics

2009

BERLIOZ Alain, TROMPETTE Philippe
Solid Mechanics using the Finite Element Method

LEMAIRE Maurice
Structural Reliability

2007

GIRARD Alain, ROY Nicolas
Structural Dynamics in Industry

GUINEBRETIÈRE René
X-ray Diffraction by Polycrystalline Materials

KRYSINSKI Tomasz, MALBURET François
Mechanical Vibrations

KUNDU Tribikram
Advanced Ultrasonic Methods for Material and Structure Inspection

SIH George C. *et al.*
Particle and Continuum Aspects of Mesomechanics

www.ingramcontent.com/pod-product-compliance
Lightning Source LLC
LaVergne TN
LVHW020421301125
826312LV00002B/27
* 9 7 8 1 7 8 6 3 0 5 1 6 9 *